T0316483

Triangular Orthogonal Functions for the Analysis of Continuous Time Systems

About the Authors

Anish Deb (b.1951) did his BTech. (1974), MTech. (1976) and PhD (Tech.) degree (1990) from the Department of Applied Physics, University of Calcutta. He started his career as a design engineer (1978) in industry and joined the Department of Applied Physics, University of Calcutta as Lecturer in 1983. In 1990 he became Reader in the same Department. Presently he is a Professor (1998). His research interests include automatic control in general and application of 'alternative' orthogonal functions in systems and control.

    Gautam Sarkar (b.1953) did his BTech. (1975), MTech. (1977) and PhD (Tech.) degree (1991) from the Department of Applied Physics, University of Calcutta. He started his career as a Research Assistant and became a Lecturer (1985) and subsequently Reader (1998) in the same Department. Presently he is in the Chair of Labonyamoyee Das Professor, which he holds since 2002. His areas of research include automatic control, fuzzy systems, microprocessor based control of electric motors, power electronics and application of piecewise constant basis functions in systems and control.

    Anindita Sengupta (b.1969) did her BTech. (1993), MTech. (1995) and PhD (Tech.) degree (2006) from the Department of Applied Physics, University of Calcutta. After gaining industrial experience for about one and half years she started her teaching career as Lecturer (1997) at North Calcutta Polytechnic. Then in 2002 she joined the Department of Electrical Engineering, Bengal Engineering and Science University as Lecturer and presently is an Assistant Professor. Currently she is engaged in research in the field of control engineering, process control and microprocessor based systems. She has published research papers in national and international journals.

Triangular Orthogonal Functions for the Analysis of Continuous Time Systems

Anish Deb
Professor
Department of Applied Physics
University of Calcutta

Gautam Sarkar
Labonyamoyee Das Professor
Department of Applied Physics
University of Calcutta

Anindita Sengupta
Assistant Professor
Department of Electrical Engineering
Bengal Engineering and Science University

ANTHEM PRESS
LONDON · NEW YORK · DELHI

Anthem Press
An imprint of Wimbledon Publishing Company
www.anthempress.com

This edition first published in UK and USA 2011
by ANTHEM PRESS
75-76 Blackfriars Road, London SE1 8HA, UK
or PO Box 9779, London SW19 7ZG, UK
and
244 Madison Ave. #116, New York, NY 10016, USA

First published in India by Elsevier 2007

British Library Cataloguing in Publication Data
A catalogue record for this book is available from the British Library.

Library of Congress Cataloging in Publication Data
A catalog record for this book has been requested.

ISBN-13: 978 0 85728 999 5 (Hbk)
ISBN-10: 0 85728 999 3 (Hbk)

This title is also available as an eBook.

To our families
for continued support, patience and understanding

Contents

Preface

It all started with Walsh functions, proposed by JL Walsh in 1922 (published in 1923). The orthonormal function set he introduced was very much dissimilar to then reigning sine-cosine functions, because it contained piecewise constant bi-valued component functions. Despite this novelty, the Walsh function attracted little attention at the time, much like its forerunner the Haar function (proposed in 1910).

However, amongst all other piecewise constant basis functions (in simple terms, staircase functions), the Walsh function suddenly became important in the mid-1960s because of its similarities, in essence, with the popular sine-cosine functions and its digital technology compatibility. This function set was a pioneer to generating the interest of researchers working in the area of communication engineering.

In the late 1980s and 1990s, orthogonal staircase functions, like the Walsh function and the block pulse function, encouraged many researchers in terms of successful applications befitting the digital age. However, the researchers, as always, kept on with zeal and vigour for better accuracy and faster computation, and this thriving attitude gave rise to many other orthogonal function sets useful for applications in the general area of systems and control. Yet compared to Walsh and block pulse functions, these new sets had to be satisfied with the back seat.

The orthogonal triangular function set is the result of such a quest, and this new function set has been applied to a few areas of control theory in this book. The audience familiar with the fundamentals of Walsh and block pulse function theory will find the material comfortable, and we hope interesting as well. For readers new to this special area, some brief introductory material has also been provided in the first few chapters, including a historical background beginning with the genesis of orthogonal staircase function sets. Overall, the book is intended for interested

readers in the academic field as well as in industry. We also hope
to generate interest in readers entering this field for the first
time. Thorough references have been given to support strongly
explorative readers.

Incidentally, the first author spent a quarter of a century with
these function sets, and the second author joined him in this field
of interest about a decade later. The third author stayed very close
to this area of research for about eight years. We felt the time
was now ripe to indulge in putting down our efforts in this field
in black and white in the form of a book, though small.

Anish Deb
Gautam Sarkar
Anindita Sengupta

Chapter 1

Walsh, Block Pulse, and Related Orthogonal Functions in Systems and Control

Orthogonal properties [1] of familiar sine–cosine functions have been known for over two centuries; but the use of such functions to solve complex analytical problems was initiated by the work of the famous mathematician Baron Jean-Baptiste-Joseph Fourier [2]. Fourier introduced the idea that an arbitrary function, even the one defined by different equations in adjacent segments of its range, could nevertheless be represented by a single analytic expression. Although this idea encountered resistance at the time, it proved to be central to many later developments in mathematics, science, and engineering.

In many areas of electrical engineering the basis for any analysis is a system of sine–cosine functions. This is mainly due to the desirable properties of frequency domain representation of a large class of functions encountered in engineering design. In the fields of circuit analysis, control theory, communication, and the analysis of stochastic problems, examples are found extensively where the completeness and orthogonal properties of such a system lead to attractive solutions. But with the application of digital techniques in these areas, awareness for other more general complete systems of orthogonal functions has developed. This "new" class of functions, though not possessing some of the desirable properties of sine–cosine functions, has other advantages to be useful in many applications in the context of digital technology. Many members of this class of orthogonal functions are piecewise constant binary valued, and therefore indicated their possible suitability in the analysis and synthesis of systems leading to piecewise constant solutions.

1.1 Orthogonal Functions and their Properties

Any time function can be synthesized completely to a tolerable degree of accuracy by using a set of orthogonal functions. For such accurate representation of a time function, the orthogonal set should be complete [1].

Let a time function $f(t)$, defined over a time interval $[0, T]$, be represented by an orthogonal function set $S_n(t)$. Then

$$f(t) = \sum_{n=0}^{\infty} c_n S_n(t) \tag{1.1}$$

where, c_n is the coefficient or weight connected to the nth member of the orthogonal set.

The members of the function set $S_n(t)$ are said to be orthogonal in the interval $0 \leq t \leq T$ if for any positive integral values of m and n, we have

$$\int_0^T S_m(t)S_n(t)dt = \delta_{mn} \text{ (a constant)} \tag{1.2}$$

where, δ_{mn} is the Kronecker delta and $\delta_{mn} = 0$ for $m \neq n$. When $m = n$ and $\delta_{mn} = 1$, then the set is said to be an orthonormal set.

An orthonormal set is said to be *complete* or *closed* if for the defined set no function can be found which is normal to each member of the set satisfying equation (1.2).

Since only a finite number of terms of the series $S_n(t)$ can be considered for practical realization of any time function $f(t)$, right-hand side (RHS) of equation (1.1) has to be truncated and we have

$$f(t) \approx \sum_{n=0}^{N} c_n S_n(t) \tag{1.3}$$

When N is large, the accuracy of representation is good enough for all practical purposes. Also, it is necessary to choose the coefficients c_n in such a manner that the mean integral squared error (MISE) is minimized. Thus,

$$\text{MISE} = \frac{1}{T} \int_0^T \left[f(t) - \sum_{n=0}^{N} c_n S_n(t) \right]^2 dt \qquad (1.4)$$

This is realized by making

$$c_n = \frac{1}{T} \int_0^T f(t) S_n(t) dt \qquad (1.5)$$

For a complete orthogonal function set, the MISE in equation (1.4) decrease monotonically to zero as N tends to infinity.

1.2 Different Types of Nonsinusoidal Orthogonal Functions

1.2.1 Haar functions

In 1910, Hungarian mathematician Alfred Haar proposed a complete set of piecewise constant binary-valued orthogonal functions that are shown in Fig. 1.1 [3,4]. In fact, Haar functions have three possible states 0 and $\pm A$ where A is a function of $\sqrt{2}$. Thus, the amplitude of the component functions varies with their place in the series.

An m-set of Haar functions may be defined mathematically in the semi-open interval $t \in [0, 1)$ as given below:

The first member of the set is

$$\text{har}(0, 0, t) = 1, \quad t \in [0, 1)$$

while the general term for other members is given by

$$\text{har}(j, n, t) = \begin{cases} 2^{j/2}, & (n-1)/2^j \le t < (n - \frac{1}{2})/2^j \\ -2^{j/2}, & (n - \frac{1}{2})/2^j \le t < n/2^j \\ 0, & \text{elsewhere} \end{cases}$$

where, $j, n,$ and m are integers governed by the relations $0 \le j \le \log_2(m), 1 \le n \le 2j$. The number of members in the set is of the form $m = 2^k$, k being a positive integer. Following the above equation, the members of the set of Haar functions can be obtained in a sequential manner. In Fig. 1.1, k is taken to be 3, thus giving $m = 8$.

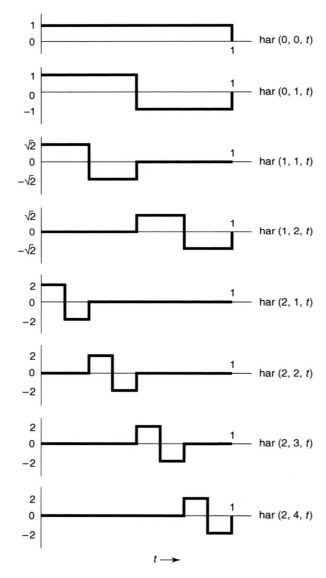

Figure 1.1. *A set of Haar functions.*

Haar's set is such that the formal expansion of given continuous function in terms of these new functions converges uniformly to the given function.

1.2.2 Rademacher functions

In 1992, inspired by Haar, German mathematician H. Rademacher presented another set of two-valued orthonormal functions [5] that are shown in Fig. 1.2. The set of Rademacher functions is orthonormal but incomplete. As seen from Fig. 1.2, the function rad(n, t) of the set is given by a square wave of unit amplitude and 2^{n-1} cycles in the semi-open interval $[0,1)$. The first member of the set rad$(0, t)$ has a constant value of unity throughout the interval.

1.2.3 Walsh functions

After the Rademacher functions were introduced in 1922, around the same time, American mathematician J.L. Walsh independently proposed yet another binary-valued complete set of normal orthogonal function Φ, later named Walsh functions [6,7], that are shown in Fig. 1.3.

As indicated by Walsh, there are many possible orthogonal function sets of this kind and several researchers, in later years, have suggested orthogonal sets [8–10] formed with the help of combinations of the well-known piecewise constant orthogonal functions.

In his original paper Walsh pointed out that, " ... Harr's set is, however, merely one of an infinity of sets which can be constructed of functions of this same character." While proposing his new set of orthonormal functions Φ, Walsh wrote " ... each function Φ takes only the values $+1$ and -1, except at a finite number of points of discontinuity, where it takes the value zero."

The Rademacher functions were found to be a true subset of the Walsh function set. The Walsh function set possesses the following properties all of which are not shared by other orthogonal functions belonging to the same class. These are:

(i) Its members are all two-valued functions.
(ii) It is a complete orthonormal set.

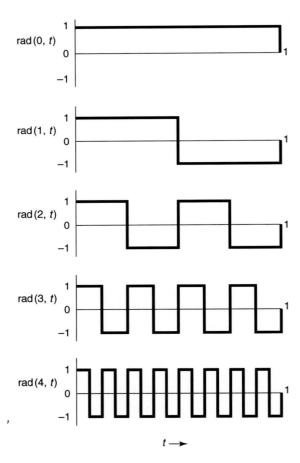

Figure 1.2. *A set of Rademacher functions.*

(iii) It has striking similarity with the sine-cosine functions, primarily with respect to their zero-crossing patterns.

1.2.4 Block pulse functions (BPF)

During the 19th century, the most important function set used for communication was block pulses. Voltage and current pulses, such as Morse code signals, were generated by mechanical switches, amplified by relays and detected by a variety of

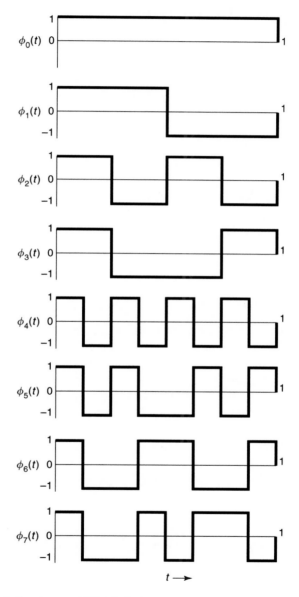

Figure 1.3. *A set of Walsh functions arranged in dyadic order.*

magnetomechanical devices. However, until recently, the set of block pulses received less attention from the mathematicians as well as application engineers possibly due to their apparent incompleteness. But disjoint and orthogonal properties of such a function set were well known. A set of block pulse functions [11–13] in the semi-open interval $t \in [0, T)$ is shown in Fig. 1.4.

An m-set block pulse function is defined as

$$\Psi_i(t) = \begin{cases} 1, & iT/m \leq t < (i+1)T/m \\ 0, & \text{elsewhere} \end{cases}$$

where, $i = 0, 1, 2, \ldots, (m-1)$.

The block pulse function set is a complete [14] orthogonal function set and can be normalized by defining the component functions in the interval $[0, T)$ as

$$\Psi_i(t) = \begin{cases} \sqrt{m}, & iT/m \leq t < (i+1)T/m \\ 0, & \text{elsewhere} \end{cases}$$

1.2.5 Slant functions

A special orthogonal function set known as the slant function set was introduced by Enomoto and Shibata [15] for image transmission analysis. These functions are also applied successfully to image processing problems [16].

Slant functions have a finite but a large number of possible states as can be seen from Fig. 1.5. The superiority of the slant function set lies in its transform characteristics, which permit a compaction of the image energy to only a few transformed samples. Thus, the efficiency of image data transmission in this form is improved. This is because the slant transform is designed to posses the following important properties [17]:

(i) constant basis vector,
(ii) slant basis vector (monotonically decreasing in constant size steps from maximum to a minimum amplitude),
(iii) sequency property,
(iv) fast computational algorithm,
(v) high-energy compaction.

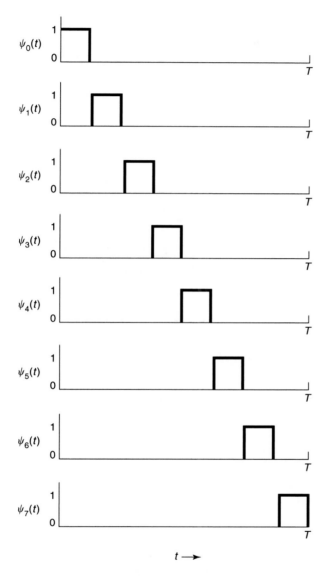

Figure 1.4. *A set of block pulse functions.*

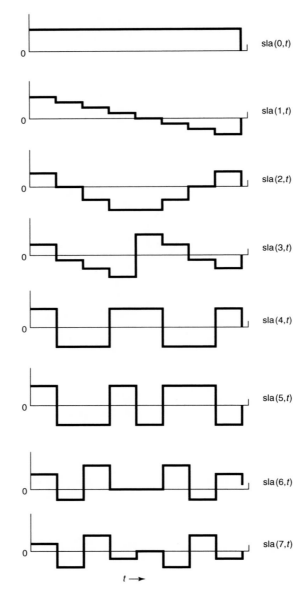

Figure 1.5. *A set of slant functions.*

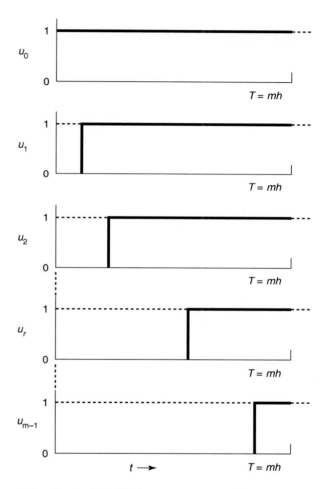

Figure 1.6. *A set of DUSF for m-component functions.*

1.2.6 Delayed unit step functions (DUSF)

Delayed unit step functions, shown in Fig. 1.6, were suggested by Hwang [18] in 1983. Though not of much use due to its dependency on BPFs, shown by Deb et al. [13], it certainly deserves to be included in the record of piecewise constant basis functions as a new variant.

$$u_i(t) = \begin{cases} 1, & t \geq ih, \quad i = 0, 1, 2, \ldots, (m-1) \\ 0, & t < ih \end{cases} \tag{1.6}$$

1.2.7 General hybrid orthogonal functions (GHOF)

So far the discussion centered on different types of orthogonal functions having a piecewise constant nature. The major departure from this class was the formulation of general hybrid orthogonal functions (GHOF) introduced by Patra and Rao [19–21]. While sine-cosine functions or orthogonal polynomials can represent a continuous function quite nicely, these functions/polynomials become unsatisfactory for approximating functions with discontinuities, jumps or dead time. For representation of such functions, undoubtedly piecewise constant orthogonal functions such as Walsh or block pulse functions, can be used more advantageously. But with functions having both continuous nature as well as a number of discontinuities in the time interval of interest, it is quite clear that none of the orthogonal functions/polynomials of continuous nature, or, for that matter, piecewise constant orthogonal functions are suitable if a reasonable degree of accuracy is to be achieved. Hence, to meet the combined features of continuity and discontinuity in such situations, the framework of GHOF proposed by Patra and Rao seemed to be more appropriate. Thus, the system of GHOF forms a hybrid basis which is both flexible and general.

Applications of GHOFs were sucessfully made by Patra and Rao in analyzing linear time invariant systems, converter fed and chopper fed dc motor drives, and in the area of self-tuning control. The main disadvantage of GHOF seems to be the required a *priori* knowledge about the discontinuities in the function, which are to be coincided with the segment boundaries of the system of GHOF to be chosen. This also requires a complex algorithm for better results.

1.2.8 Variants of block pulse functions

In 1995, a pulse-width modulated version of the block pulse function set was presented by Deb et al. [22,23] where, the pulse-width of the component functions of the BPF set was gradually increased (or, decreased) depending upon the nature of the square integrable function to be handled.

In 1998, a further variant of the BPF set was proposed by Deb et al. [24] where, the set was called sample and hold function (SHF) set and the same was utilized for the analysis of sampled data systems with zero-order hold.

1.3 Walsh Functions in Systems and Control

Among all the orthogonal functions outlined earlier in the chapter, Walsh function based analysis first became more attractive to the researchers from 1962 onwards [7,25–27]. The reason for such success was mainly due to its binary nature. One immediate advantage is the task of analog multiplication. To multiply any signal by a Walsh function, the problem reduces to an appropriate sequence of sign changes, which makes this usually difficult operation both simple and potentially accurate [25]. However, in system analysis, Walsh functions were employed during early 1970s. As a consequence, the advantages of Walsh analysis were unraveled to the workers in the field compared to the use of conventional sine-cosine functions. Ultimately, the mathematical groundwork of the Walsh analysis became strong to lure interested researchers to try every new application based upon this function.

In 1973, it was Corrington [28] who proposed a new technique for solving linear as well as nonlinear differential and integral equations with the help of Walsh functions. In 1975, important technical papers relating Walsh functions to the field of systems and control were published. New ideas were proposed by Rao [29–35,37] and Chen [38–43]. Other notable workers were Le Van et al. [44], Tzafestas [45], Chen [46–49], Mahapatra [50], Paraskevopoulos [51], Moulden [52], Deb and Datta [53–55], Lewis [56], Marszalek [57], Dai and Sinha [58], Deb et al. [59–62], and others.

The first positive step for the development of the Walsh domain analysis was the formulation of the operational matrix for integration. This was done independently by Rao [29], Chen [38], and Le Van et al. [44]. Le Van sensed that since the integral operator matrix had an inverse, the inverse must be the differential operator in the Walsh domain. However, he could not represent the general form of the operator matrix that was done by Chen [38,39]. Interestingly, the operational matrix for integration was first presented by Corrington [28] in the form of a table. But he failed to recognize the potentiality of the table as a matrix. This was first pointed out by Chen and he presented Walsh domain analysis with the operational matrices for integration as well as differentiation:

   (i)   to solve the problems of linear systems by state space model [38];
  (ii)   to design piecewise constant gains for optimal control [39];
 (iii)   to solve optimal control problem [40];
 (iv)   in variational problems [41];
  (v)   for time domain synthesis [42];
 (vi)   for fractional calculus as applied to distributed systems [43].

   Rao used Walsh functions for:

   (i)   system identification [29];
  (ii)   optimal control of time-delay systems [31];
 (iii)   identification of time-lag systems [32];
 (iv)   transfer function matrix identification [34] and piecewise linear system identification [35];
  (v)   parameter estimation [36];
 (vi)   solving functional differential equations and related problems [37].

   He first formulated operational matrices for stretch and delay [37]. He proposed a new technique for extension of computation beyond the limit of initial normal interval with the help of "single term Walsh series" approach [33], and estimated the error due to the use of different operational matrices [36]. Rao and Tzafestas indicated the potentiality of Walsh and related functions in the area of systems and control in a review paper [63]. Tzafestas [64]

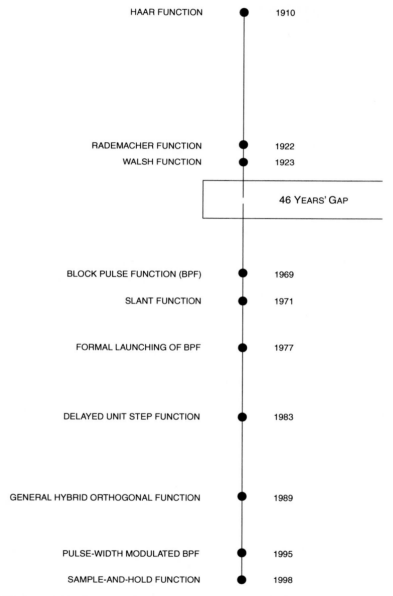

Figure 1.7. *Time scale history of piecewise constant and related basis function family.*

assessed the role of Walsh functions in signal and system analysis and design in a rich collection of papers.

W. L. Chen defined a "shift Walsh matrix" for solving delay-differential equations [47], and used Walsh functions for parameter estimation of bilinear systems [46] as well as in the analysis of multi-delay systems [49]. Paraskevopoulos determined the transfer functions of a single input single output (SISO) system from its impulse response with the help of Walsh functions and a fast Walsh algorithm [51]. Tzafestas applied Walsh series approach for lumped and distributed system identification [45]. Mahapatra used Walsh functions for solving matrix Riccati equation arising in optimal control studies of linear diffusion equations [50]. Moulden's work was concerned with the application of Walsh spectral analysis of ordinary differential equations in a very formal manner [52]. Deb applied Walsh functions to analyze power-electronic systems [53]. Deb and Datta was the first to define the Walsh Operational Transfer Function (WOTF) for the analysis of linear SISO systems [53, 59–61]. Deb was the first to notice the oscillatory behavior in the Walsh domain analysis of first-order systems [62].

1.4 Block Pulse Functions in Systems and Control

The earliest work concerning completeness and suitability of BPF for use in place of Walsh functions, is a small technical note of Rao and Srinivasan [65]. Later Kwong and Chen [14], Chen and Lee [66], and Sloss and Blyth [67] discussed convergence properties of BPF series and the BPF solution of a linear time invariant system.

Sannuti's paper [68] on the analysis and synthesis of dynamical systems in state space was a significant step toward BPF applications. Shieh et al. [69] dealt with the same problems. The doctoral dissertation of Srinivasan [70] contained several applications of BPF to a variety of problems. Rao and Srinivasan proposed methods of analysis and synthesis for delay systems [71] where an operational matrix for delay via BPF was proposed.

Chen and Jeng [72] considered systems with piecewise constant delays. BPF are also used to invert numerically Laplace transforms [73–76]. Differential equations related to the dynamics of current collection mechanism of electric locomotives contain terms with a stretched argument. Such equations have been treated in Ref. [77] using BPF. Chen [78] also dealt with scaled systems. BPFs have been used in obtaining discrete-time approximations of continuous-time systems. Shieh et al. [69] and recently Sinha et al. [79] gave some interesting results in this connection. The BPF method of discretization has been compared with other techniques employing bilinear transformation, state transition matrix, etc.

The higher powers of the operational matrix for integration accomplished the task of repeated integration. However, the use of higher powers led to accumulation of errors at each stage of integration. This has been recognized by Rao and Palanisamy who gave one shot operational matrices for repeated integration via BPF and Walsh functions. Wang [80] deals with the same aspect suggesting improvements in operational matrices for fractional and operational calculus. Palanisamy reveals certain interesting aspects of the operational matrix for integration. Optimal controls for time-invariant systems and time-varying systems have been worked out [82]. Kawaji [83] gave an analysis of linear systems with observers.

Replacement of Walsh function by block pulse took place in system identification algorithms for computational advantage. Shih and Chia [84] used BPF in identifying delay systems. Jan and Wong [85] and Cheng and Hsu [86] identified bilinear models. Multidimensional BPFs have been proposed by Rao and Srinivasan [87]. These were used in solving partial differential equations. Nath and Lee [88] discussed multidimensional extensions of block pulse with applications.

Identification of nonlinear distributed systems and linear feedback systems via block pulse functions were done by Hsu and Cheng [89] and Kwong and Chen [90]. Palanisamy and Bhattacharya also used block pulse functions in system identification [91] and in analyzing stiff systems [92]. Solution of multipoint boundary value problems and integral equations were obtained using a set of BPF [93,94]. In parameter estimation of bilinear systems Cheng and Hsu [95] applied block pulse functions. Still many more applications of block pulse functions remain to be mentioned.

Thus, block pulse function continued to reign over other piecewise constant orthogonal functions with its simple but powerful attributes.

1.5 Conclusion

With this rich background, it seems worthwhile to explore this field further. Here, in this work, piecewise linear orthogonal triangular functions, derived from block pulse functions, are proposed and applied to different areas of continuous time control systems.

References

1. Sansone, G., *Orthogonal functions*, Interscience Publishers Inc., New York, 1959.
2. Fourier, J. B., *Théorie analytique de la Chaleur*, 1828. *The analytic theory of heat*, 1878, (English edition) Reprinted by Dover Pub. Co, New York, 1955.
3. Haar, Alfred, Zur theorie der orthogonalen funktionen systeme, Math. Annalen, vol. **69**. pp. 331–371, 1910.
4. Beauchamp, K. G., *Walsh functions and their applications*, Academic Press, London, 1975.
5. Rademacher, H., Einige sätze von allegemeinen orthogonal funktionen, Math. Annalen, vol. **87**, pp. 122–138, 1922.
6. Walsh, J. L., A closed set of normal orthogonal functions, Amer. J. Math., vol. **45**, pp. 5–24, 1923.
7. Harmuth, H. F., *Transmission of information by orthogonal functions* (2nd Ed.), Springer-Verlag, Berlin, 1972.
8. Fino, B. J. and Algazi, V. R., Slant-Haar transform, Proc. IEEE, vol. **62**, pp. 653–654, 1974.
9. Rao, K. R., Narasimhan, M. A., and Revuluri, K., Image data processing by Hadamard–Haar transform, IEEE Trans Computers, vol. **C-24**, pp. 888–896, 1975.

10. Huang, D. M., Walsh–Hadamard–Haar hybrid transform, IEEE Proc. 5th Int. Conf. on Pattern Recognition, pp. 180–182, 1980.
11. Wu, T. T., Chen, C. F., and Tsay Y. T., Walsh operational matrices for fractional calculus and their application to distributed systems, IEEE Symposium on Circuits and Systems, Munich, Germany, April, 1976.
12. Jiang, J. H. and Schaufelberger, W., *Block pulse functions and their application in control system*, LNCIS–179, Springer-Verlag, Berlin, 1992.
13. Deb, Anish, Sarkar, G. and Sen, S. K., Block pulse functions, the most fundamental of all piecewise constant basis functions, Int. J. Syst. Sci., vol. **25**, no. 2, pp. 351–363, 1994.
14. Kwong, C. P. and Chen, C. F., The convergence properties of block pulse series, Int. J. Syst. Sci., vol. **12**, no. 6, pp. 745–751, 1981.
15. Enomoto, H. and Shibata, K., Orthogonal transform coding system for television signals, Proc. Symp. Appl. of Walsh functions, Washington DC, USA, pp. 11–17, 1971.
16. Pratt, W. K., Welch, L. R. and Chen, W., Slant transform for image coding, Proc. Symp. Appl. of Walsh functions, Washington DC, USA, pp. 229–234, March 1972.
17. Pratt, W. K., *Digital image processing*, John Wiley & Sons, New York, 1978.
18. Hwang, Chyi, Solution of functional differential equation via delayed unit step functions, Int. J. Syst. Sci., vol. **14**, no. 9, pp. 1065–1073, 1983.
19. Patra, A. and Rao, C. P., General hybrid orthogonal functions— a new tool for the analysis of power-electronic systems, IEEE Trans. Industrial Electronics, vol. **36**, no. 3, pp. 413– 424, 1989.
20. Patra, A. and Rao, G. P., General hybrid orthogonal functions and some potential applications in systems and control, Proc. IEE, Part D, Control Theory and Appl., vol. **136**, no. 4, pp. 157–163, 1989.
21. Patra, A. and Rao, G. P., Continuous-time approach to self-tuning control: Algorithm, implementation and assessment, Proc. IEE, Part D, Control Theory and Appl., vol. **136**, no. 6, pp. 333–340, 1989.
22. Deb, A., Sarkar, G. and Sen, S. K., Linearly pulse-width modulated block pulse functions and their application to linear SISO feedback control system identification, Proc. IEE, Part D, Control Theory and Appl., vol. **142**, no. 1, pp. 44–50, 1995.

23. Deb, A., Sarkar, G. and Sen, S. K., A new set of pulse width modulated generalised block pulse functions (PWM-GBPF) and their application to cross/auto-correlation of time varying functions, Int. J. Syst. Sci., vol. **26**, no. 1, pp. 65–89, 1995.
24. Deb, Anish, Sarkar, Gautam, Bhattacharjee, Manabrata, and Sen, Sunit, K., A new set of piecewise constant orthogonal functions for the analysis of linear SISO systems with sample-and-hold, J. Franklin Instt., vol. **335B**, no. 2, pp. 333–358, 1998.
25. Beauchamp, K. C., *Walsh and related functions and their applications*, Academic Press, London, 1984.
26. Maqusi, M., *Applied Walsh analysis*, Heyden, London, 1981.
27. Hammond, J. L. and Johnson, R. S., A review of orthogonal square wave functions and their application to linear networks, J. Franklin Instt., vol. **273**, pp. 211–225, March, 1962.
28. Corrington, M. S., Solution of differential and integral equations with Walsh functions, IEEE Trans. Circuit Theory, vol. **CT-20**, no. 5, pp. 470–476, 1973.
29. Rao, G. P. and Sivakumar, L., System identification via Walsh functions, Proc. IEE, vol. **122**, no. 10, pp. 1160–1161, 1975.
30. Rao, G. P. and Palanisamy, K. R., A new operational matrix for delay via Walsh functions and some aspects of its algebra and applications, 5th National Systems Conference, NSC-78, PAU Ludhiana (India), September, pp. 60–61, 1978.
31. Rao, G. P. and Palanisamy, K. R., Optimal control of time-delay systems via Walsh functions, 9th IFIP Conference on optimisation techniques, Polish Academy of Sc., Syst. Research Instt., Poland, September, 1979.
32. Rao, G. P. and Sivakumar, L., Identification of time-lag systems via Walsh functions, IEEE Trans. Automatic Control, vol. **AC-24**, no. 5, pp. 806–808, 1979.
33. Rao, G. P., Palanisamy, K. R. and Srinivasan, T., Extension of computation beyond the limit of initial normal interval in Walsh series analysis of dynamical systems, IEEE Trans. Automatic Control, vol. **AC-25**, no. 2, pp. 317–319, 1980.
34. Rao, G. P. and Sivakumar, L., Transfer function matrix identification in MIMO systems via Walsh functions, Proc. IEEE, vol. **69**, no. 4, pp. 465–466, 1981.
35. Rao, G. P. and Sivakumar, L., Piecewise linear system identification via Walsh functions, Int. J. Syst. Sci., vol. **13**, no. 5, pp. 525–530, 1982.

36. Rao, G. P., *Piecewise constant orthogonal functions and their applications in systems and control*, LNCIS–55, Springer-Verlag, Berlin, 1983.
37. Rao, G. P. and Palanisamy, K. R., Walsh stretch matrices and functional differential equations, IEEE Trans. Automatic Control, vol. **AC-2** 1, no. 1, pp. 272–276, 1982.
38. Chen, C. F. and Hsiao, C. H., A state space approach to Walsh series solution of linear systems, Int. J. Syst. Sci., vol. **6**, no. 9, pp. 833–858, 1975.
39. Chen, C. F. and Hsiao, C. H., Design of piecewise constant gains for optimal control via Walsh functions, IEEE Trans. Automatic Control, vol. **AC-20**, no. 5, pp. 596–603, 1975.
40. Chen, C. F. and Hsiao, C. H., Walsh series analysis in optimal control, Int. J. Control, vol. **21**, no. 6, pp. 881–897, 1975.
41. Chen, C. F. and Hsiao, C. H., A Walsh series direct method for solving variational problems, J. Franklin Inst., vol. **300**, no. 4, pp. 265–280, 1975.
42. Chen, C. F. and Hsiao, C. H., Time domain synthesis Via Walsh functions, IEE Proceedings, vol. **122**, no. 5, pp. 565–570, 1975.
43. Chen, C. F., Tsay, Y. T. and Wu, T. T., Walsh operational matrices for fractional calculus and their application to distributed systems, J. Franklin Instt., vol. **303**, no. 3, pp. 267–284, 1977.
44. Van, T. Le, Tam, L. D. C. and Houtte, N. Van, On direct algebraic solutions of linear differential equations using Walsh transforms, IEEE Trans. Circuits and Syst., vol. **CAS-22**, no. 5, pp. 419–422, 1975.
45. Tzafestas, S. G., Walsh series approach to lumped and distributed system identification, J. Franklin Instt., vol. **305**, no. 4, pp. 199–220, 1978.
46. Chen, W. L. and Shih, Y. P., Parameter estimation of bilinear systems via Walsh functions, J. Franklin Instt., vol. **305**, no. 5, pp. 249–257, 1978.
47. Chen, W. L. and Shih, Y. P., Shift Walsh matrix and delay differential equations, IEEE Trans. Automatic Control, vol. **AC-23**, no. 6, pp. 1023–1028, 1978.
48. Chen, W. L. and Lee, C. L., Walsh series expansions of composite functions and its applications to linear systems, Int. J. Syst. Sci., vol. **13**, no. 2, pp. 219–226, 1982.
49. Chen, W. L., Walsh series analysis of multi-delay systems, J. Franklin Inst., vol. **303**, no. 4, pp. 207–217, 1982.

50. Mahapatra, G. B. Solution of optimal control problem of linear diffusion equation via Walsh functions, IEEE Trans. Automatic Control, vol. **AC-25**, no. 2, pp. 319–321, 1980.
51. Paraskevopoulos, P. N. and Varoufakis, S. J., Transfer function determination from impulse response via Walsh functions, Int. J. Circuit Theory and Appl., vol. **8**, no. pp. 85–89, 1980.
52. Moulden, T. H. and Scott, M. A., Walsh spectral analysis for ordinary differential equations: Part 1–Initial value problem, IEEE Trans. Circuits and Syst., vol. **CAS-35**, no. 6, pp. 742–745, 1988.
53. Deb, Anish, and Datta, A. K., Time response of pulse-fed SISO systems using Walsh operational matrices, Advances in Modelling and Simulation, AMSE press, vol. **8**, no. 2, pp. 30–37, 1987.
54. Deb, Anish, On Walsh domain analysis of power-electronic systems, Ph.D. (Tech.) dissertation, University of Calcutta, 1989.
55. Deb, Anish, and Datta, A. K., On analytical techniques of power electronic circuit analysis, IETE Tech. Review, vol. **7**, no. 1, pp. 25–32, 1990.
56. Lewis, F. L., Mertzios, B. G., Vachtsevanos, G. and Christodoulou, M. A. Analysis of bilinear systems using Walsh functions, IEEE Trans. Automatic Control, vol. **35**, no. 1, pp. 119–123, 1990.
57. Marszalek W., Orthogonal functions analysis of singular systems with impulsive responses, Proc. IEE, Part D, Control Theory and Appl., vol. **137**, no. 2, pp. 84–86, 1990.
58. Dai, H. and Sinha N. K., Robust coefficient estimation of Walsh functions, Proc. IEE, Part D, Control Theory and Appl., vol. **137**, no. 6, pp. 357–363, 1990.
59. Deb, Anish and Datta, A. K., Analysis of continuously variable pulse-width modulated system via Walsh functions, Int. J. Sci., vol. **23**, no. 2, pp. 151–166, 1992.
60. Deb, Anish and Datta A. K., Analysis of pulse-fed power electronic circuits using Walsh functions, Int. J. Electronics, vol. **62**, no. 3, pp. 449–459, 1987.
61. Deb, A., Sarkar G., Sen S. K. and Datta, A. K., A new method of analysis of chopper-fed DC series motor using Walsh function, Proc. 4th European Conf. on Power Electronics and Applications (EPE '91), Florence, Italy, 1991.
62. Deb, Anish and Fountain, D. W., A note on oscillations in Walsh domain analysis of first order systems, IEEE Trans. Circuts and Syst., vol. **CAS-38**, no. 8, pp. 945–948, 1991.

63. Rao, G. P. and Tzafestas, S. G.. A decade of piecewise constant orthogonal functions in systems and control, Math. and Computers in Simulation, vol. **27**, no. 5 & 6, pp. 389–407, 1985.
64. Tzafestas, S. G. (Ed.), *Walsh functions in signal and systems analysis and design*, Van Nostrand Renhold Co., New York, 1985.
65. Rao, G. P. and Srinivasan, T., Remarks on "Author's reply" to "Comments on design of piecewise constant gains for optimal control via Walsh functions". IEEE Trans. Automatic Control, vol. **AC-23**, no. 4, pp. 762–763, 1978.
66. Chen, W. L. and Lee, C. L., On the convergence of the block-pulse series solution of a linear time-invariant system, Int. J. Syst. Sci., vol. **13**, no. 5, pp. 491–498, 1982.
67. Sloss, G. and Blyth, W. F., A priori error estimates for Corrington's Walsh function method, J. Franklin Instt., vol. **331B**, no. 3, pp. 273–283, 1994.
68. Sannuti, P., Analysis and synthesis of dynamic systems via block pulse functions, Proc. IEE. vol. **124**, no. 6, pp. 569–571, 1977.
69. Shieh, L. A., Yates, R. E. and Navarro, J. M., Representation of continuous time state equations by discrete-time state equations, IEEE Trans. Signal, Man and Cybernatics, vol. **SMC-8**, no. 6, pp. 485–492, 1978.
70. Srinivasan, T., Analysis of dynamical systems via block-pulse functions, Ph.D. dissertation, Dept. of Electrical Engineering, I. I. T., Kharagpur, 721302, India, 1979.
71. Rao, G. P. and Srinivasan, T., Analysis and synthesis of dynamic systems containing time delays via block-pulse functions, Proc. IEE, vol. **125**, no. 9, pp. 1064–1068, 1978.
72. Chen, W. L. and Jeng, B. S., Analysis of piecewise constant delay systems via block-pulse functions, Int. J. Syst. Sci., vol. **12**, no. 5, pp. 625–633, 1981.
73. Hwang, C., Guo, T. Y. and Shih, Y. P., Numerical inversion of multidimensional Laplace transforms via block-pulse function, Proc. IEE, Part D, Control Theory Appl., vol. **130**, no. 5, pp. 250–254, 1983.
74. Marszalek, W., The block-pulse function method of the two-dimensional Laplace transform, Int. J. Syst. Sci., vol. **14**, no. 11, pp. 1311–1317, 1983.
75. Jiang, Z. H., New approximation method for inverse Laplace transforms using block-pulse functions, Int. J. Syst. Sci., vol. **18**, no. 10, pp. 1873–1888, 1987.

76. Shieh, L. A. and Yates, R. E., Solving inverse Laplace transform, linear and nonlinear state equations using block-pulse functions, Computers and Elect. Engg, vol. **6**, pp. 3–17, 1979.
77. Rao, G. P. and Srinivasan, T., An optimal method of solving differential equations characterizing the dynamic of a current collection system for an electric locomotive, J. Inst. Maths. Appl., vol. **25**, no. 4, pp. 329–342, 1980.
78. Chen, W. L., Block-pulse series analysis of scaled systems, Int. J. Syst. Sci., vol. **12**, no. 7, pp. 885–891, 1981.
79. Sinha, N. K. and Zhou, Qijie, Discrete-time approximation of multivariable continuous-time systems, Proc. IEE, Part D, Control Theory and Appl., vol. **130**, no. 3, pp. 103–110, 1983.
80. Wang, Chi-Hsu, On the generalisation of block pulse operational matrices for fractional and operational calculus, J. Franklin Instt., vol. **315**, no. 2, pp. 91–102, 1983.
81. Palanisamy, K. R., A note on block-pulse function operational matrix for integration, Int. J. Syst. Sci., vol. **14**, no. 11, pp. 1287–1290, 1983.
82. Hsu Ning-Show and Cheng Bing, Analysis and optimal control of time-varying linear systems via block-pulse functions, Int. J. Control, vol. **33**, no. 6, pp. 1107–1122, 1981.
83. Kawaji, S., Block-pulse series analysis of linear systems incorporating observers, Int. J. Control, vol. **37**, no. 5, pp. 1113–1120, 1983.
84. Shih, Y. P. and Chia, W. K., Parameter estimation of delay systems via block- pulse functions, J. Dynamic Syst., Measurement and Control, vol. **102**, no. 3, pp. 159–162, 1980.
85. Jan, Y. G. and Wong, K. M., Bilinear system identification by block pulse functions, J. Franklin Instt., vol. **512**, no. 5, pp. 349–359, 1981.
86. Cheng Bing and Hsu Ning-Show, Analysis and parameter estimation of bilinear systems via block-pulse functions, Int. J. Control, vol. **36**, no. 1, pp. 53–65, 1982.
87. Rao, G. P. and Srinivasan, T., Multidimensional block pulse functions and their use in the study of distributed parameter systems, Int. J. Syst. Sci., vol. **11**, no. 6, pp. 689–708, 1980.
88. Nath, A. K. and Lee, T. T., On the multidimensional extension of block-pulse functions and their applications, Int. J. Syst. Sci., vol. **14**, no. 2, pp. 201–208, 1983.

89. Hsu, Ning-Show and Cheng, Bing, Identification of nonlinear distributed systems via block-pulse functions, Int. J. Control, vol. **36**, no. 2, pp. 281–291, 1982.
90. Kwong, C. P. and Chen, C. F., Linear feedback systems identification via block-pulse functions, Int. J. Syst. Sci., vol. **12**, no. 5, pp. 635–642, 1981.
91. Palanisamy, K. R. and Bhattacharya, D. K., System identification via block-pulse functions, Int. J. Syst. Sci., vol. **12**, no. 5, pp. 643–647, 1981.
92. Palanisamy, K. R. and Bhattacharya, D. K., Analysis of stiff systems via single step method of block-pulse functions, Int. J. Syst. Sci., vol. **13**, no. 9, pp. 961–968, 1982.
93. Kalat, J. and Paraskevopoulos, P. N., Solution of multipoint boundary value problems via block pulse functions, J. Franklin Instt., vol. **324**, no. 1, pp. 73–81, 1987.
94. Kung, F. C. and Chen, S. Y., Solution of integral equations using a set of block pulse functions, J. Franklin Instt., vol. **306**, no. 4, pp. 283–291, 1978.
95. Cheng, B. and Hsu, N. S., Analysis and parameter estimation of bilinear systems via block pulse functions, Int. J. Control, vol. **36**, no. 1, pp. 53–65, 1982.

Chapter 2

A Newly Proposed Triangular Function Set and Its Properties

Walsh and block pulse functions (BPFs) approximate time functions in a piecewise constant manner. In this chapter, we propose a complementary pair of triangular function (TF) sets for approximating time functions in a piecewise linear manner and discuss their various properties. We start with a brief review of Walsh and related functions, and their respective operational matrices for integer integration/differentiation.

2.1 Walsh Functions and Related Operational Matrix for Integration

It is known that Walsh functions $\Phi(t)$ are a set of square waves, which are orthonormal [1]. Figure 1.3 shows the functions from ϕ_0 to ϕ_7 in the dyadic order.

For any arbitrary function $f(t)$, we can expand it into Walsh series, if the function is absolutely integrable in $[0, 1)$. That is

$$f(t) \approx \sum_{n=0}^{N} c_n \phi_n(t) \tag{2.1}$$

where

$$c_n = \int_0^1 \phi_n(t) f(t) \mathrm{d}t \tag{2.2}$$

are determined such that the following integral squared error $\in$ is minimized, i.e.,

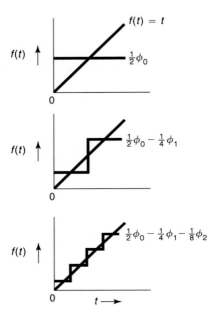

Figure 2.1. *Decomposing a ramp function into Walsh functions.*

$$\epsilon = \int_0^1 \left[f(t) - \sum_{n=0}^{N} c_n \phi_n(t) \right]^2 dt \qquad (2.3)$$

For example, if we expand $f(t) = t$, the result is [2]

$$f(t) = \frac{1}{2}\phi_0(t) - \frac{1}{4}\phi_1(t) - \frac{1}{8}\phi_2(t) - \frac{1}{16}\phi_4(t) + \cdots \qquad (2.4)$$

RHS of equations (2.4) forms a staircase wave that approximates the original function t, as shown in Fig. 2.1.

If we perform integration on the Walsh functions, we get various triangular waves. Evaluating the Walsh co-efficients for these triangular waves and arranging properly, we have the following relation

$$
\begin{bmatrix}
\int \phi_0 dt \\
\int \phi_1 dt \\
\int \phi_2 dt \\
\int \phi_3 dt \\
\int \phi_4 dt \\
\int \phi_5 dt \\
\int \phi_6 dt \\
\int \phi_7 dt
\end{bmatrix}
=
\begin{bmatrix}
\frac{1}{2} & -\frac{1}{4} & -\frac{1}{8} & 0 & -\frac{1}{16} & 0 & 0 & 0 \\
\frac{1}{4} & 0 & 0 & -\frac{1}{8} & 0 & -\frac{1}{16} & 0 & 0 \\
\frac{1}{8} & 0 & 0 & 0 & 0 & 0 & -\frac{1}{16} & 0 \\
0 & \frac{1}{8} & 0 & 0 & 0 & 0 & 0 & -\frac{1}{16} \\
\frac{1}{16} & 0 & 0 & 0 & 0 & 0 & 0 & 0 \\
0 & \frac{1}{16} & 0 & 0 & 0 & 0 & 0 & 0 \\
0 & 0 & \frac{1}{16} & 0 & 0 & 0 & 0 & 0 \\
0 & 0 & 0 & \frac{1}{16} & 0 & 0 & 0 & 0
\end{bmatrix}
\begin{bmatrix}
\phi_0 \\
\phi_1 \\
\phi_2 \\
\phi_3 \\
\phi_4 \\
\phi_5 \\
\phi_6 \\
\phi_7
\end{bmatrix}
$$

$$(2.5)$$

or

$$
\int \Phi_{(\mathbf{8})} dt = G_{(\mathbf{8}\times\mathbf{8})} \Phi_{(\mathbf{8})} \tag{2.6}
$$

Here with no loss of generality, only the first eight Walsh waves are considered. Subscripts indicate the dimension taken. It is preferable to take 2^p component Walsh functions, where p is an integer. Note that the ith row of the square matrix shows the Walsh series coefficients of the function $\int \phi_{i-1} dt$, where $i = 1, 2, \ldots, 8$. For example, the first row of equations (2.5) gives the Walsh coefficients of $\int \phi_0 dt$. These eight coefficients may be found in the first eight terms of equation (2.4).

The square matrix G of equation (2.6) performs like an integrator and is called the Walsh operational matrix for integer integration. It is interesting to note that if G is partitioned into four equal parts as shown, the upper left corner of $G_{(8\times8)}$ is identical to $G_{(4\times4)}$; and the upper left part of partitioned $G_{(4\times4)}$ is $G_{(2\times2)}$. Therefore, the regularity of the structure of G matrix enabled researchers [2] to enlarge it to higher dimensions.

We write the general operational matrix for integration as follows:

$$
\mathbf{G}_{(m \times m)} =
\begin{bmatrix}
\frac{1}{2} & & -\frac{2}{m}\mathbf{I}_{(\frac{m}{8})} & -\frac{1}{m}\mathbf{I}_{(\frac{m}{4})} & \\
\frac{2}{m}\mathbf{I}_{(\frac{m}{8})} & \mathbf{O}_{(\frac{m}{8})} & & & -\frac{1}{2m}\mathbf{I}_{(\frac{m}{2})} \\
\frac{1}{m}\mathbf{I}_{(\frac{m}{4})} & & \mathbf{O}_{(\frac{m}{4})} & & \\
\frac{1}{2m}\mathbf{I}_{(\frac{m}{2})} & & & \mathbf{O}_{(\frac{m}{2})} &
\end{bmatrix}
$$

$$(2.7)$$

The operational matrix of equation (2.6) clearly performs the integration for integer calculus. Therefore, it can be used to solve linear differential equations effectively. Chen and Hsiao demonstrated its power in solving state equations [2], in synthesizing transfer functions in time domain [3], and in finding time-varying gains for optimal control [4].

2.2 BPFs and Related Operational Matrices

BPFs were derived from Walsh functions by Chen et al. [5]. The set of BPFs is more fundamental that the Walsh function set [6]. BPFs are related to Walsh functions by the following formula:

$$
\begin{bmatrix}
\phi_0 \\ \phi_1 \\ \phi_2 \\ \phi_3
\end{bmatrix}
=
\begin{bmatrix}
1 & 1 & 1 & 1 \\
1 & 1 & -1 & -1 \\
1 & -1 & 1 & -1 \\
1 & -1 & -1 & 1
\end{bmatrix}
\begin{bmatrix}
\psi_0 \\ \psi_1 \\ \psi_2 \\ \psi_3
\end{bmatrix}
$$

$$(2.8)$$

or, in general,

$$\mathbf{\Phi}_{(\mathbf{m})} \approx \mathbf{W}_{(\mathbf{m} \times \mathbf{m})} \mathbf{\Psi}_{(\mathbf{m})} \tag{2.9}$$

where, $\mathbf{\Psi}_{(\mathbf{m})}$ is the m-set BPFs with unity height and $1/m$ width.

The square matrix in equation (2.8), denoted by $\mathbf{W}_{(\mathbf{4} \times \mathbf{4})}$, is called the Walsh matrix. The relation between this matrix and Walsh waveforms is shown in Fig. 2.2. From Fig. 2.2, we see that it is very easy to construct the Walsh matrix because the

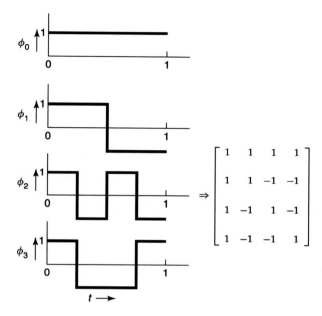

Figure 2.2. *First four Walsh functions and the Walsh matrix.*

elements of the matrix correspond directly to the magnitude of the waveforms.

One of the properties of **W** matrix is that

$$\mathbf{W}^2_{(\mathbf{m}\times\mathbf{m})} = m\mathbf{I_m} \tag{2.10}$$

or

$$\mathbf{W}^{-1}_{(\mathbf{m}\times\mathbf{m})} = \frac{1}{m}\mathbf{W}_{(\mathbf{m}\times\mathbf{m})} \tag{2.11}$$

Substituting equation (2.9) into equation (2.6), we get,

$$\int_0^t \mathbf{W}_{(\mathbf{m}\times\mathbf{m})}\mathbf{\Psi}_{(\mathbf{m})}\mathrm{d}t \approx \mathbf{G}_{(\mathbf{m}\times\mathbf{m})}\mathbf{W}_{(\mathbf{m}\times\mathbf{m})}\mathbf{\Psi}_{(\mathbf{m})} \tag{2.12}$$

Therefore,

$$\int_0^t \mathbf{\Psi}_{(\mathbf{m})}\mathrm{d}t = \mathbf{W}^{-1}_{(\mathbf{m}\times\mathbf{m})}\mathbf{G}_{(\mathbf{m}\times\mathbf{m})}\mathbf{W}_{(\mathbf{m}\times\mathbf{m})}\mathbf{\Psi}_{(\mathbf{m})} \tag{2.13}$$

We define the following:

$$\mathbf{W}^{-1}_{(\mathbf{m}\times\mathbf{m})}\mathbf{G}_{(\mathbf{m}\times\mathbf{m})}\mathbf{W}_{(\mathbf{m}\times\mathbf{m})} \triangleq \mathbf{P}_{(\mathbf{m}\times\mathbf{m})} \tag{2.14}$$

From equation (2.11) we have

$$\mathbf{P}_{(\mathbf{m}\times\mathbf{m})} = \frac{1}{m}\mathbf{W}\mathbf{G}\mathbf{W} \tag{2.15}$$

where, we have dropped the subscript ($\mathbf{m} \times \mathbf{m}$) on RHS for simplicity.

Evaluation of the above similarity transformation yields

$$\mathbf{P}_{(\mathbf{m}\times\mathbf{m})} = \frac{1}{m}\begin{bmatrix} \frac{1}{2} & 1 & 1 & 1 & \cdots & 1 \\ 0 & \frac{1}{2} & 1 & 1 & \cdots & 1 \\ 0 & 0 & \frac{1}{2} & 1 & \cdots & 1 \\ . & . & . & . & \cdots & . \\ . & . & . & . & \cdots & . \\ . & . & . & . & \cdots & . \\ 0 & 0 & 0 & 0 & \cdots & \frac{1}{2} \end{bmatrix}_{(m\times m)} \tag{2.16}$$

Matrix $\mathbf{P}$ is the operational matrix for integration in the BPF domain. Of course, this matrix can be directly obtained from integration of Fig. 2.3(a), by reading the average values of the magnitudes in four subintervals of $\frac{1}{4}$ second duration each, as shown in Fig. 2.3(b).

As already mentioned, the set of BPFs is more fundamental than Walsh functions. Also, the block pulse domain operational matrix for integration shown in equation (2.16) is simpler than the Walsh domain operational matrix.

The $\mathbf{P}$ matrix may also be expressed in the following form:

$$\mathbf{P}_{(\mathbf{m}\times\mathbf{m})} = \frac{1}{m}\left(\frac{1}{2}\mathbf{I}_{(\mathbf{m})} + \mathbf{Q}_{(\mathbf{m}\times\mathbf{m})} + \mathbf{Q}^2_{(\mathbf{m}\times\mathbf{m})} + \cdots + \mathbf{Q}^{m-1}_{(\mathbf{m}\times\mathbf{m})}\right)$$

$$= \frac{1}{2}[\mathbf{I}_{(\mathbf{m})} + \mathbf{Q}_{(\mathbf{m}\times\mathbf{m})}][\mathbf{I}_{(\mathbf{m})} - \mathbf{Q}_{(\mathbf{m}\times\mathbf{m})}]^{-1} \tag{2.17}$$

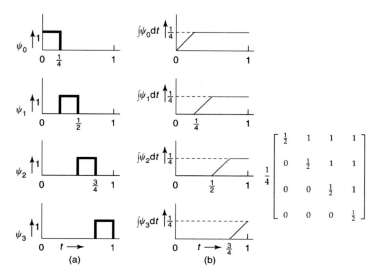

Figure 2.3. *BPFs and their first integrations.*

where, $\mathbf{Q}_{(m \times m)}$ is the delay matrix [5] given by

$$\mathbf{Q}_{(m \times m)} = \begin{bmatrix} 0 & 1 & 0 & 0 & \cdots & 0 \\ 0 & 0 & 1 & 0 & \cdots & 0 \\ 0 & 0 & 0 & 1 & \cdots & 0 \\ \cdot & \cdot & \cdot & \cdot & \cdots & \cdot \\ \cdot & \cdot & \cdot & \cdot & \cdots & \cdot \\ 0 & 0 & 0 & 0 & \cdots & 0 \end{bmatrix} \triangleq \begin{bmatrix} 0 & & & \\ \cdot & & \mathbf{I}_{(m-1)} & \\ \cdot & & & \\ 0 & & & \\ \hline 0 & 0 & \cdots & 0 \end{bmatrix}$$

(2.18)

From equation (2.17), the operational matrix for differentiation, $\mathbf{D}_{(m \times m)}$, may be obtained by inverting the RHS. Thus

$$\mathbf{D}_{(m \times m)} = 2 \big[\mathbf{I}_{(m)} - \mathbf{Q}_{(m \times m)} \big] \big[\mathbf{I}_{(m)} + \mathbf{Q}_{(m \times m)} \big]^{-1}$$

A set of BPF $\mathbf{\Psi}_{(m)}(t)$ containing m component functions in the semi-open interval $[0, T)$ is given by

$$\mathbf{\Psi}_{(m)}(t) \triangleq \big[\psi_0(t) \quad \psi_1(t) \quad \cdots \quad \psi_i(t) \quad \cdots \quad \psi_{m-1}(t) \big]^{\mathrm{T}}$$

(2.19)

where, $[\ldots]^{\mathrm{T}}$ denotes transpose.

The ith component $\psi_i(t)$ of the BPF vector $\mathbf{\Psi_{(m)}(t)}$ is defined as

$$\psi_i(t) = \begin{cases} 1 & iT/m \leq t < (i+1)T/m \\ 0 & \text{otherwise} \end{cases}$$

where, $i = 0, 1, 2, \ldots, (m-1)$.

A square integrable time function $f(t)$ of Lebesgue measure may be expanded into an m-term BPF series in $t \in [0, T)$ as

$$f(t) \approx [c_0 \quad c_1 \quad c_2 \quad \ldots \quad c_i \quad \ldots \quad c_{m-1}]\mathbf{\Psi_{(m)}(t)} \triangleq \mathbf{C^T \Psi_{(m)}(t)}$$

$$(2.20)$$

The constant coefficients c_i's in equation (2.20) are given by [6]

$$c_i = \frac{1}{h} \int_{ih}^{(i+1)h} f(t)\mathrm{d}t \tag{2.21}$$

where, $h = T/m$ is the duration of each component BPF along the time scale.

In the m-term BPF domain, over the interval $[0, T)$, following equation (2.16), the operational matrix for integration $\mathbf{P_{(m)}}$ is the upper triangular matrix:

$$\mathbf{P_{(m)}} \triangleq h \left[\!\!\left[\frac{1}{2} \quad 1 \quad \ldots \quad \ldots \quad \ldots \quad 1 \right]\!\!\right]_{m \times m} \tag{2.22}$$

where

$$[a \quad b \quad c] \triangleq \begin{bmatrix} a & b & c \\ 0 & a & b \\ 0 & 0 & a \end{bmatrix}$$

The matrix $\mathbf{P_{(m)}}$ performs as an integrator in the BPF domain and it is pivotal in any BPF domain analysis. It is to be noted that, for deriving $\mathbf{P_{(m)}}$, time interval considered is $[0, T)$, instead of $[0, 1)$, with m component functions. Hence, the operational matrix of equation (2.16) is multiplied by T to obtain equation (2.22).

Thus, approximate integration of a function $f(t)$, using equations (2.20) and (2.22), is

$$\int f(t)dt \approx \mathbf{C}^{\mathsf{T}}\mathbf{P}\boldsymbol{\Psi}_{(m)}(t) \qquad (2.23)$$

where, $\mathbf{C}$, $\mathbf{P}$, $\boldsymbol{\Psi}_{(m)}(t)$ are of order m.

Generalized operational matrices for multiple integration [7,8] were also computed by researchers for application in systems and control.

When the square integrable function $f(t)$ is not available in its analytic form, but is represented graphically, or by tabulated data, it is apparent that equation (2.21) cannot be used. If the function $f(t)$ is available as equidistant samples $s_0, s_1, s_2, \ldots, s_m$, it is obvious that the average value of the function in the kth subinterval is given by

$$f_k = \frac{s_k + s_{(k+1)}}{2}$$

The average values f_k's can be considered as coefficients of *nonoptimal BPFs* where the coefficients are not computed via equation (2.21), but derived from two consecutive samples of the function $f(t)$ available in discrete form.

Another kind of nonoptimal BPF is the sample-and-hold function (SHF), which is discussed in the following section.

2.3 Sample-and-Hold Functions [9]

From the nature of BPFs, it is easy to see that these functions resemble the time response of a zero-order hold (ZOH) device. Truly speaking, the first member of the BPF set, ψ_0, cannot be distinguished from the time response of a ZOH. In view of this resemblance, a new set of PCBF termed as sample-and-hold functions (SHF) was introduced by Deb et al. for the analysis of control system with sample-and-hold. It should be noted that, representation of a square integrable function $f(t)$ in the semi-open interval $[0, T)$ by means of SHF is quite different from that of BPF representation.

The function $f(t)$ may be represented by the sample-and-hold technique in the interval $[0, T)$ by considering

$$f_i(t) \approx f(ih), \quad i = 0, 1, 2, \ldots, (m-1) \qquad (2.24)$$

where, h is the sampling period $(= T/m)$, $f_i(t)$ is the amplitude of the function $f(t)$ at time $ih \le t < (i+1)h$, $f(ih)$ is the first term of the Taylor series expansion of the function $f(t)$ around the point $t = ih$, because, for a ZOH the amplitude of the function $f(t)$ at $t = ih$ is held constant for the duration h.

SHF are similar to BPFs in many aspects. A set of SHF, comprised of m component functions, is defined as

$$z_i(t) = \begin{cases} 1 & ih \le t < (i+1)h \\ 0 & \text{otherwise} \end{cases} \qquad (2.25)$$

where, $i = 0, 1, 2, \ldots, (m-1)$.

Considering the nature of SHF, which is a look alike of the BPF set, it is easy to conclude that this set is orthogonal as well as complete in $t \in [0, T)$ like the BPFs. However, the special property of the SHF is revealed by using the sample-and-hold concept in the derivation of the required operational matrices.

If a time signal $f(t)$ is fed to an S/H device as shown in Fig. 2.4, the output of the device approximates $f(t)$ as per equation (2.24). Thus

$$f(t) \approx \sum_{i=0}^{m-1} f_i z_i(t) = [f_0 \quad f_1 \quad \cdots \quad f_{m-1}] \, \mathbf{Z_{(m)}(t)}$$

$$\text{for } i = 0, 1, 2, \ldots, (m-1)$$

$$\triangleq \mathbf{F_{(m)}^T Z_{(m)}(t)} \qquad (2.26)$$

where, we have chosen an m-set SHF.

Unlike the coefficient c_i's of BPFs, the coefficient f_i's of equation (2.26) are given by

$$\int f_i = f(t)\delta(t - ih)\mathrm{d}t, \quad ih \le T \qquad (2.27)$$

where, $\delta(t)$ is the well-known Dirac delta function.

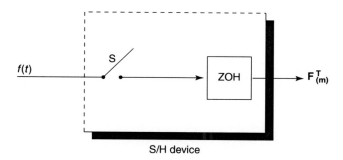

S/H device

Figure 2.4. *Sample-and-hold device.*

2.3.1 SHF operational matrix for integration

As with the BPFs, the operational matrix for integration in the SHF domain follows the relation

$$\int \mathbf{Z}_{(m)}(\mathbf{t})dt \approx \mathbf{P1}_{(m)} \cdot \mathbf{Z}_{(m)}(\mathbf{t}) \tag{2.28}$$

where, $\mathbf{Z}_{(m)}(\mathbf{t})$ is the m-term SHF vector and $\mathbf{P1}_{(m)}$ is the operational matrix for integration of order m in the SHF domain given by [9]

$$\mathbf{P1}_{(m)} \triangleq h\,[0 \quad 1 \quad 1 \quad 1 \quad 1 \quad \ldots \quad 1]_{m \times m}$$

It is noted that, like equation (2.17), $\mathbf{P1}_{(m)}$ can be expressed as

$$\mathbf{P1}_{(m)} = h\mathbf{Q}_{(m)}[\mathbf{I}_{(m)} - \mathbf{Q}_{(m)}]^{-1} \tag{2.29}$$

$\mathbf{P1}_{(m)}$ is defective in the sense that it is not invertible like $\mathbf{P}_{(m)}$ and it has an eigenvalue 0 repeated m times.

2.4 From BPF to a Newly Defined Complementary Set of Triangular Functions

Though BPFs are effective for analysis and synthesis of various control systems, it is not baseless hunch to think that a staircase solution provided by the BPF domain analysis may introduce relatively more error than an equivalent piecewise linear solution.

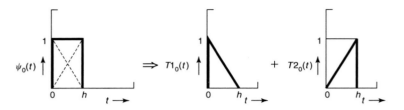

Figure 2.5. *Dissection of BPF into two triangular functions.*

In the following, we have dissected a BPF (say, the first member $\psi_0(t)$ of an m-set BPF) into two triangular functions as shown in Fig. 2.5.

Thus, we have

$$\psi_0(t) = T1_0(t) + T2_0(t)$$

For the whole set of BPF, $\mathbf{\Psi_m(t)}$, we can thus generate two sets of orthogonal TF, namely $\mathbf{T1_{(m)}(t)}$ and $\mathbf{T2_{(m)}(t)}$ such that

$$\mathbf{\Psi_{(m)}(t)} = \mathbf{T1_{(m)}(t)} + \mathbf{T2_{(m)}(t)} \qquad (2.30)$$

Figures 2.6(a) and (b) show the orthogonal triangular function sets, $\mathbf{T1_{(m)}(t)}$ and $\mathbf{T2_{(m)}(t)}$, where m has been chosen arbitrarily as 8. It could be said that these two sets are complementary to each other as far as BPFs are considered. For convenience, we refer $\mathbf{T1_{(m)}(t)}$ as the left-handed triangular function (LHTF) vector and $\mathbf{T2_{(m)}(t)}$ as the right-handed triangular function (RHTF) vector.

Now, following equation (2.19), we can express the m-set triangular function [10] vectors as

$$\mathbf{T1_{(m)}(t)} \triangleq [T1_0(t) \ T1_1(t) \ T1_2(t) \ \dots \ T1_i(t) \ \dots \ T1_{m-1}(t)]^T$$

and

$$\mathbf{T2_{(m)}(t)} \triangleq [T2_0(t) \ T2_1(t) \ T2_2(t) \ \dots \ T2_i(t) \ \dots \ T2_{m-1}(t)]^T$$

where, $[\dots]^T$ denotes transpose.

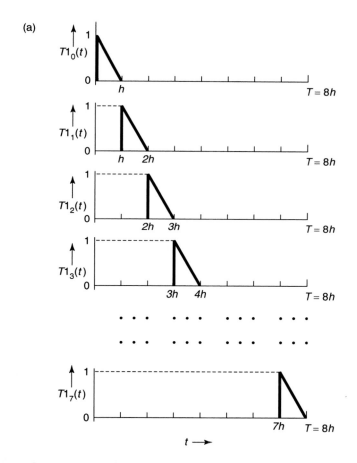

Figure 2.6. *(a) A set of LHTF* $\mathbf{T1_{(m)}}(t)$, $m = 8$.

The *i*th component of the LHTF vector $\mathbf{T1_{(m)}}(\mathbf{t})$ is defined as

$$T1_i(t) = \begin{cases} 1 - (t - ih)/h & \text{for } ih \leq t < (i+1)h \\ 0 & \text{otherwise} \end{cases} \quad (2.31)$$

and the *i*th component of the RHTF vector $\mathbf{T2_{(m)}}(\mathbf{t})$ is defined as

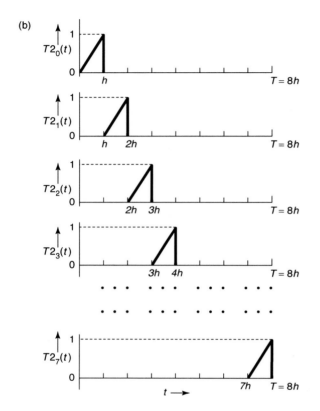

Figure 2.6. *(b) A set of RHTF* **$T2_{(m)}(t)$**, $m = 8$.

$$T2_i(t) = \begin{cases} (t - ih)/h & \text{for } ih \leq t < (i+1)h \\ 0 & \text{otherwise} \end{cases} \quad (2.32)$$

where, $i = 0, 1, 2, \ldots, (m - 1)$.

2.5 Piecewise Linear Approximation of a Square Integrable Function *f(t)*

Let us consider a square integrable function $f(t)$ shown in Fig. 2.7. We take a few sample points of this function $f(t)$ and connect these points by straight lines to form a piecewise linear

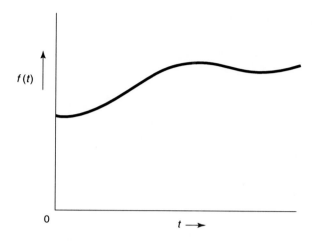

Figure 2.7. A square integrable function $f(t)$.

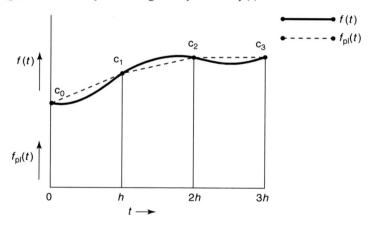

Figure 2.8. The function $f(t)$ and its piecewise linear approximation $f_{pl}(t)$ in $0 \leq t < 3h$ using four equidistant sample points.

version of $f(t)$. We call this as transformed function $f_{pl}(t)$. For
the function $f(t)$ shown in Fig. 2.7, we take four (say) equidistant
sample points from $t = 0$ to $3h$ and the transformed piecewise
linear version of $f(t)$ is shown in Fig. 2.8.

For convenience, the sample points are chosen to be equidistant, the sampling period being h seconds. Thus, $f(0) = c_0$,

$f(h) = c_1$, $f(2h) = c_2$, and $f(3h) = c_3$. It is obvious that, if h is smaller—or, in other words, the number of samples per second is higher—the piecewise linear function $f_{pl}(t)$ will approach its continuous version $f(t)$.

In Fig. 2.8, we have three trapeziums placed side by side which approximate the function $f(t)$ from $t = 0$ up to the point $t = 3h$. These three trapeziums may be decomposed into two triangles each, in all six triangles, as shown in Fig. 2.9.

It is apparent that, these six triangles, when added graphically—or, mathematically for that matter—yield the piecewise linear function $f_{pl}(t)$. The sample values c_0, c_1, etc. become the altitudes of different triangles for such piecewise linear approximation.

In mathematical terms, the function $f(t) \approx f_{pl}(t)$ is given by

$$f(t) \approx f_{pl}(t) = \{c_0 T1_0(t) + c_1 T2_0(t)\} + \{c_1 T1_1(t) + c_2 T2_1(t)\}$$
$$+ \{c_2 T1_2(t) + c_3 T2_2(t)\}, \quad 0 \le t < 3h$$

$$= [c_0\ c_1\ c_2] \begin{bmatrix} T1_0(t) \\ T1_1(t) \\ T1_2(t) \end{bmatrix} + [c_1\ c_2\ c_3] \begin{bmatrix} T2_0(t) \\ T2_1(t) \\ T2_2(t) \end{bmatrix}$$

$$(2.33)$$

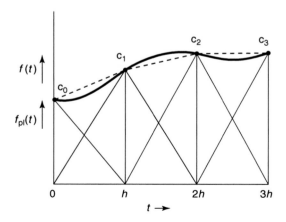

Figure 2.9. *Three trapeziums or six triangles add up to form $f_{pl}(t)$, the approximate piecewise linear version of $f(t)$.*

It is of interest to note that, the sample values of $f(t)$—namely, c_0, c_1, etc—have been used as the weights in equation (2.33) like in sine-cosine expansion in the well-known Fourier series. The weights change the altitudes of the triangular basis functions given by equations (2.31) and (2.32). Since the base of each basis function remains fixed $(= h)$, the slope of the hypotenuse changes to suit the piecewise linear reconstruction of $f(t)$.

In general, following equation (2.33) a square integrable time function $f(t)$ of Lebesgue measure may be expanded into an m-term TF series in $t \in [0, T)$ as

$$
\begin{aligned}
f(t) \approx {} & [c_0 \quad c_1 \quad c_2 \quad \ldots \quad c_i \quad \ldots c_{m-1}]\mathbf{T1}_{(m)}(t) \\
& + [d_0 \quad d_1 \quad d_2 \quad \ldots \quad d_i \quad \ldots \quad d_{m-1}]\mathbf{T2}_{(m)}(t) \\
\triangleq {} & \mathbf{C}^T\mathbf{T1}_{(m)}(t) + \mathbf{D}^T\mathbf{T2}_{(m)}(t)
\end{aligned}
\tag{2.34}
$$

The constant coefficients c_i's and d_i's in equation (2.34) are given by

$$
c_i \triangleq f(ih) \quad \text{and} \quad d_i \triangleq f[(i+1)h]
\tag{2.35}
$$

The following relation between c_i's and d_i's are also noted.

$$
c_{i+1} = d_i
$$

It may be noted that, the coefficients c_i's and d_i's in equation (2.35) are *not optimal* in the sense that they do not help in approximating $f(t)$ in a least square manner. This would essentially imply that the mean integral squared error (MISE) is not *minimized*. However, the advantage of choosing c_i's and d_i's as different samples of $f(t)$ for obtaining a piecewise linear solution, instead of conventional integration formula, is obvious. It is apparent from equations (2.21) and (2.35) that unlike BPFs, the TF representation does not need any integration to evaluate the coefficients, thereby reducing a lot of computational burden. In a sense, we can call these c_i's and d_i's of equation (2.35) to be *nonoptimal*.

Similar nonoptimal coefficients were used by Deb et al. [9] for analyzing discrete-data control systems with sample-and-hold. Had we used optimal coefficients, the set of SHF [9] would have

Table 2.1. Basic properties of BPFs and TFs

Property	BPF	TF
Piecewise constant	Yes	No
Orthogonal	Yes	Yes
Finite	Yes	Yes
Disjoint	Yes	Yes
Orthonormal	Can easily be normalized	Can easily be normalized
Implementation	Easily Implementable	Easily Implementable
Coefficient determination of $f(t)$	Involves integration of $f(t)$ and scaling	Needs only samples of $f(t)$
Accuracy of analysis	Provides staircase solution with more error than TF analysis	Provides piecewise linear solution having less error than BPF analysis

been converted to well-known BPF set. This would have led to the loss of advantages of using the SHF (actually, nonoptimal BPF) set for analyzing systems with sample-and-hold. Obviously, *the advantages were derived at the cost of nonminimum MISE.*

Derivation of coefficients for optimal representation via triangular functions is given in Section 2.8.

Basic properties of BPFs and TFs are tabulated in Table 2.1 to provide a qualitative appraisal.

2.6 Orthogonality of Triangular Basis Functions

The condition of orthogonality for the LHTF set $\mathbf{T1}_{(m)}(\mathbf{t})$ demands that

$$\int_0^T T1_p(t)T1_q(t)\mathrm{d}t = \delta_{pq}$$

where, δ_{pq} is the Kronecker delta given by

$$\delta_{pq} = 0, \quad \text{for } p \neq q \quad \text{and} \quad \delta_{pq} = \text{constant, for } p = q$$

Since the members $T1_i(t)$, $i = 0, 1, 2, \ldots, (m-1)$, of the LHTF set are mutually disjoint, the product

$$T1_p(t)T1_q(t) = 0 \quad \text{for } p \neq q$$

Hence, $\int_0^T T1_p(t)T1_q(t)dt = 0$, for $p \neq q$
When $p = q$, we have

$$\int_0^T [T1_p(t)]^2 dt = \int_{ph}^{(p+1)h} [T1_p(t)]^2 dt \qquad (2.36)$$

The function $T1_p(t)$ is given by

$$T1_p(t) = u(t - ph) - \frac{(t - ph)}{h}u(t - ph)$$
$$+ \frac{t - (p + 1)h}{h}u\{t - (p + 1)h\}$$

Substituting $T1_p(t)$ in equation (2.36), we have

$$\int_{ph}^{(p+1)h} [T1_p(t)]^2 dt = \int_{ph}^{(p+1)h} \left[u(t - ph) - \frac{(t - ph)}{h}u(t - ph) \right.$$
$$\left. + \frac{\{t - (p + 1)h\}}{h}u\{t - (p + 1)h\} \right]^2 dt$$
$$= \int_{ph}^{(p+1)h} \left[1 + \frac{(t - ph)^2}{h^2} - 2\frac{(t - ph)}{h} \right] dt$$
$$= \frac{h}{3}(\text{constant})$$

In a similar manner, the condition of orthogonality of the RHTF set $\mathbf{T2_{(m)}(t)}$ may be established. Since, the members of this set are mutually disjoint like the LHTF set $\mathbf{T1_{(m)}(t)}$, we have

$$\int_0^T T2_p(t)T2_q(t)dt = 0, \quad \text{for } p \neq q$$

When $p = q$, we have

$$\int_0^T [T2_p(t)]^2 dt = \int_{ph}^{(p+1)h} [T2_p(t)]^2 dt \qquad (2.37)$$

The function $T2_p(t)$ is given by

$$T2_p(t) = \frac{(t - ph)}{h} u(t - ph)$$
$$- \frac{t - (p + 1)h}{h} u\{t - (p + 1)h\} - u\{t - (p + 1)h\}$$

Substituting $T2_p(t)$ in equation (2.37), we get

$$\int_{ph}^{(p+1)h} [T2_p(t)]^2 dt = \int_{ph}^{(p+1)h} \left[\frac{(t - ph)}{h} u(t - ph) \right.$$
$$- \frac{\{t - (p + 1)h\}}{h} u\{t - (p + 1)h\}$$
$$\left. - u\{t - (p + 1)\}h \right]^2 dt$$
$$= \int_{ph}^{(p+1)h} \frac{(t - ph)^2}{h^2} dt$$
$$= \frac{h}{3}(\text{constant})$$

This proves the orthogonality of the triangular function sets. It is noted that, like BPFs, the triangular function sets can easily be normalized.

2.7 A Few Properties of Orthogonal TF

In this section, we deal with a few basic properties of the orthogonal triangular function sets.

Lemma 1
The product of two triangular functions $T1_i$ and $T1_j (T1_i, T1_j \in$ **T1**$_{(m)}$, and $i, j \le m$, in the semi-open interval $[0, T)$, expressed in triangular function domain, is given by

$$T1_i T1_j = \begin{cases} 0 & \text{for } i \ne j \\ T1_i & \text{for } i = j \end{cases}$$

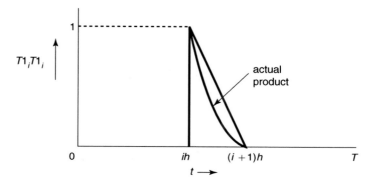

Figure 2.10. *The product $T1_iT1_i$ and its triangular function representation.*

Proof: Since the component triangular functions $T1_i$, $i = 0, 1, 2, \ldots, m - 1$ are mutually disjoint, it follows that

$$T1_iT1_j = 0 \quad \text{for } i \neq j$$

When $i = j$, the product at the sample points, namely ih and $(i + 1)h$ are 1 and 0, respectively. Since we are concerned only with the triangular function representation of the product, shown in Fig. 2.10, the results is $T1_i$ only. Thus

$$T1_iT1_j = T1_i \quad \text{for } i = j \quad \blacksquare$$

Lemma 2

The product of two triangular functions $T2_i$ and $T2_j (T2_i, T2_j \in$ **T2**$_{(m)}$, and $i, j \leq m$, in the semi-open interval $[0, T)$, expressed in triangular function domain, is given by

$$T2_iT2_j = \begin{cases} 0 & \text{for } i \neq j \\ T2_i & \text{for } i = j \end{cases}$$

Proof: Since the component triangular functions $T2_i$, $i = 0, 1, 2, \ldots, (m - 1)$ are mutually disjoint, it follows that

$$T2_iT2_j = 0 \quad \text{for } i \neq j$$

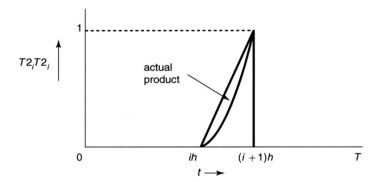

Figure 2.11. *The product $T2_iT2_j$ and its triangular function representation.*

When $i = j$, the product at the sample points, namely ih and $(i + 1)h$ are 0 and 1, respectively. Since we are concerned only with the triangular function representation of the product, shown in Fig. 2.11, the results is $T2_i$ only. Thus

$$T2_iT2_j = T2_i \quad \text{for } i = j \quad \blacksquare$$

Lemma 3
The product of two triangular functions $T1_i$ and $T2_j$ ($T1_i \in$ **T1**$_{(m)}$, $T2_j \in$ **T2**$_{(m)}$ and $i, j \le m$, in the semi-open interval $[0, T)$, expressed in triangular function domain, is given by

$$T1_iT2_j = 0 \quad \text{for all } i, j \le m$$

Proof: Since the component triangular functions $T1_i$, $i = 0, 1, 2, \ldots, (m - 1)$ and $T2_j$, $j = 0, 1, 2, \ldots, (m - 1)$ are mutually disjoint, it follows that

$$T1_iT2_j = 0 \quad \text{for } i \ne j$$

When $i = j$, the product at the sample points, namely ih and $(i + 1)h$ are both 0. Since we are concerned only with the

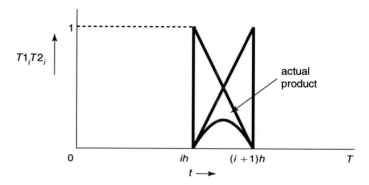

Figure 2.12. *The product $T1_i T2_i$ and its triangular function representation.*

triangular function representation of the product, shown in Fig. 2.12, the results is zero. Hence

$$T1_i T2_j = 0 \quad \text{for } i = j$$

Thus

$$T1_i T2_j = 0 \quad \text{for all } i, j \leq m \quad \blacksquare$$

Theorem 1

If two square integrable functions $f(t)$ and $g(t)$ are expanded via an m-set triangular functions as $(\mathbf{A}^T\mathbf{T1} + \mathbf{B}^T\mathbf{T2})$ and $(\mathbf{C}^T\mathbf{T1} + \mathbf{D}^T\mathbf{T2})$, respectively, then the triangular function expansion of the product $h(t) = f(t)g(t)$ is expressed as $(\mathbf{A}^T * \mathbf{C}^T\mathbf{T1} + \mathbf{B}^T * \mathbf{D}^T\mathbf{T2})$.

Here $\mathbf{A}^T * \mathbf{C}^T$ means the product matrix contains elements equal to the product of corresponding elements of the component matrices $\mathbf{A}^T$ and $\mathbf{C}^T$.

Proof: Since the functions $f(t)$ and $g(t)$ are square integrable, their product $h(t) = f(t)g(t)$ is integrable [11]. Also, from the proposition

$$h(t) = f(t)g(t) \approx (\mathbf{A}^T\mathbf{T1} + \mathbf{B}^T\mathbf{T2})(\mathbf{C}^T\mathbf{T1} + \mathbf{D}^T\mathbf{T2})$$

By virtue of *Lemmas 1–3*, we have

$$h(t) = f(t)g(t) \approx (\mathbf{A}^\mathrm{T} * \mathbf{C}^\mathrm{T}\mathbf{T1} + \mathbf{B}^\mathrm{T} * \mathbf{D}^\mathrm{T}\mathbf{T2}) \quad \blacksquare$$

Note: If the elements of $\mathbf{A}^\mathrm{T}$ are denoted by a_i and those of $\mathbf{C}^\mathrm{T}$ by c_i, then the elements of $\mathbf{A}^\mathrm{T} * \mathbf{C}^\mathrm{T}$ are denoted by $a_i c_i$. Alternatively,

$$\begin{aligned}
\mathbf{A}^\mathrm{T} * \mathbf{C}^\mathrm{T} &= \mathbf{A}^\mathrm{T}\mathrm{diag}[c_0 \quad c_1 \quad c_2 \quad \cdots \quad c_{(m-2)} \quad c_{(m-1)}] \\
&= \mathbf{C}^\mathrm{T}\mathrm{diag}[a_0 \quad a_1 \quad a_2 \quad \cdots \quad a_{(m-2)} \quad a_{(m-1)}]
\end{aligned}$$
$$(2.38)$$

Calling $\mathrm{diag}[p_0 \quad p_1 \quad p_2 \quad \cdots \quad p_{(m-2)} \quad p_{(m-1)}] = \mathbf{P_{(m \times m)}}$, equation (2.38) may be written as

$$\mathbf{A}^\mathrm{T} * \mathbf{C}^\mathrm{T} = \mathbf{A}^\mathrm{T}\mathbf{c_{(m \times m)}} = \mathbf{C}^\mathrm{T}\mathbf{a_{(m \times m)}} \qquad (2.39)$$

Remark: Theorem 1 implies that, if $h(t) = f(t)g(t)$, then given $h(t) = \mathbf{E}^\mathrm{T}\mathbf{T1} + \mathbf{F}^\mathrm{T}\mathbf{T2}$ and (either of $f(t)$ or $g(t)$—say, $f(t)$) $f(t) = (\mathbf{A}^\mathrm{T}\mathbf{T1} + \mathbf{B}^\mathrm{T}\mathbf{T2})$, $g(t)$ is then given by $g(t) = (\mathbf{E}^\mathrm{T} \wedge \mathbf{A}^\mathrm{T}\mathbf{T1} + \mathbf{F}^\mathrm{T} \wedge \mathbf{B}^\mathrm{T}\mathbf{T2})$.

Here $\mathbf{E}^\mathrm{T} \wedge \mathbf{A}^\mathrm{T}$ is a matrix whose elements are (e_i/a_i), formed with e_i and a_i, where e_i's are the elements of $\mathbf{E}^\mathrm{T}$ and a_i's are the elements of the matrix $\mathbf{A}^\mathrm{T}$. Alternatively,

$$\mathbf{E}^\mathrm{T} \wedge \mathbf{A}^\mathrm{T} = \mathbf{E}^\mathrm{T} \, \mathrm{diag}\left[\frac{1}{a_0} \quad \frac{1}{a_1} \quad \frac{1}{a_2} \quad \cdots \quad \frac{1}{a_{(m-2)}} \quad \frac{1}{a_{(m-1)}}\right]$$

2.7.1 Integration of the inner products of the basis vectors T1 and T2

The product of two basis vectors **T1(t)** with **T1(t)**$^\mathrm{T}$ is

$$\begin{aligned}
\mathbf{T1(t)T1(t)}^\mathrm{T} &= [T1_0 \quad T1_1 \quad T1_2 \quad \cdots \quad T1_{(m-1)}]^\mathrm{T} \\
&\times [T1_0 \quad T1_1 \quad T1_2 \quad \cdots \quad T1_{(m-1)}] \quad (2.40)
\end{aligned}$$

That is

$$\mathbf{T1(t)T1(t)}^{\mathrm{T}}$$

$$= \begin{bmatrix} T1_0 \\ T1_1 \\ T1_2 \\ \ldots \\ T1_i \\ \ldots \\ T1_{(m-1)} \end{bmatrix} \begin{bmatrix} T1_0 & T1_1 & T1_2 & \ldots & T1_i & \ldots & T1_{(m-1)} \end{bmatrix}$$

$$= \begin{bmatrix} T1_0^2 & T1_0T1_1 & T1_0T1_2 & \ldots \\ T1_1T1_0 & T1_1^2 & T1_1T1_2 & \ldots \\ \ldots & \ldots & \ldots & \ldots \\ T1_iT1_0 & T1_iT1_1 & T1_iT1_2 & \ldots \\ \ldots & \ldots & \ldots & \ldots \\ T1_{(m-1)}T1_0 & T1_{(m-1)}T1_1 & & \ldots \end{bmatrix}$$

$$\begin{bmatrix} T1_0T1_i & \ldots & T1_0T1_{(m-1)} \\ T1_1T1_i & \ldots & T1_1T1_{(m-1)} \\ \ldots & \ldots & \ldots \\ T1_i^2 & \ldots & T1_iT1_{(m-1)} \\ \ldots & \ldots & \ldots \\ T1_{(m-1)}T1_i & \ldots & T1_{(m-1)}^2 \end{bmatrix}$$

By virtue of Lemma 1, we can write

$$\mathbf{T1(t)T1(t)}^{\mathrm{T}}$$

$$= \mathrm{diag}\begin{bmatrix} T1_0^2 & T1_1^2 & T1_2^2 & \ldots & T1_i^2 & \ldots & T1_{(m-1)}^2 \end{bmatrix}$$

$$= \mathrm{diag}\begin{bmatrix} T1_0 & T1_1 & T1_2 & \ldots & T1_i & \ldots & T1_{(m-1)} \end{bmatrix}$$

$$\tag{2.41}$$

Integration of equation (2.41) yields

$$\int_a^b \mathbf{T1(t)T1(t)}^{\mathrm{T}}\mathrm{d}t$$

$$= \int_a^b \mathrm{diag}\begin{bmatrix} T1_0 & T1_1 & T1_2 & \ldots & T1_{(m-1)} \end{bmatrix}\mathrm{d}t \tag{2.42}$$

Let us call

$$\text{diag}\left[\int T1_0 dt \int T1_1 dt \int T1_2 dt \dots \int T1_i dt \dots \int T1_{(m-1)} dt\right]$$
$$= \boldsymbol{\Phi1} \text{ (say)}$$

Then, for the definite integral of equation (2.42), the result may be written as

$$\int_a^b \mathbf{T1(t)T1(t)}^\mathsf{T} dt = [\boldsymbol{\Phi1}]_a^b = \boldsymbol{\Phi1(b)} - \boldsymbol{\Phi1(a)} = \boldsymbol{\Phi1'} \text{ (say)}$$
$$(2.43)$$

where, $\boldsymbol{\Phi1(a)}$ and $\boldsymbol{\Phi1(b)}$ are obviously $m \times m$ matrices.

If the limits of integration are chosen as $a = 0$ and $b = mh$, then equation (2.42) becomes

$$\int_0^{mh} \mathbf{T1(t)T1(t)}^\mathsf{T} dt = \boldsymbol{\Phi1(mh)} - \boldsymbol{\Phi1(0)}$$

$$= \text{diag}\left[\underbrace{\frac{h}{2} \quad \frac{h}{2} \quad \frac{h}{2} \quad \cdots \quad \frac{h}{2}}_{m\text{-terms}}\right] \quad (2.44)$$

Since, from Fig. 2.6(a), $\int_0^{mh} T1_i dt$ gives the area under the function $T1_i$ and this area is equal to $\frac{h}{2}$.

Similarly, the product of two basis vectors $\mathbf{T2}$ with $\mathbf{T2}^\mathsf{T}$ is

$$\mathbf{T2(t)T2(t)}^\mathsf{T} = \begin{bmatrix} T2_0 & T2_1 & T2_2 & \cdots & T2_{(m-1)} \end{bmatrix}^\mathsf{T}$$
$$\times \begin{bmatrix} T2_0 & T2_1 & T2_2 & \cdots & T2_{(m-1)} \end{bmatrix}$$

By using *Lemma 2*, we can write

$$\mathbf{T2(t)T2(t)}^\mathsf{T} = \text{diag}\begin{bmatrix} T2_0 & T2_1 & T2_2 & \cdots & T2_{(m-1)} \end{bmatrix}$$
$$(2.45)$$

Integration of equation (2.45) yields

$$\int_a^b \mathbf{T2(t)T2(t)}^T dt = \mathbf{\Phi2(b)} - \mathbf{\Phi2(a)} = \mathbf{\Phi2}' \text{ (say)} \qquad (2.46)$$

Similar to equation (2.44), for $a = 0$ and $b = mh$, equation (2.46) becomes

$$\int_0^{mh} \mathbf{T2(t)T2(t)}^T dt = \mathbf{\Phi2(mh)} - \mathbf{\Phi2(0)}$$

$$= \text{diag} \left[\underbrace{\frac{h}{2} \quad \frac{h}{2} \quad \frac{h}{2} \quad \cdots \quad \frac{h}{2}}_{m\text{-terms}} \right] \qquad (2.47)$$

By virtue of *Lemma 3*, it is apparent that

$$\int_a^b \mathbf{T1(t)T2(t)}^T dt = \int_a^b \mathbf{T2(t)T1(t)}^T dt = 0 \qquad (2.48)$$

2.8 Function Approximation via Optimal Triangular Function Coefficients

Earlier we have developed the theoretical basis for the proposed triangular function set where the expansion coefficients were chosen as sample values of the function to be approximated. However, conventional orthogonal analysis demands that the coefficients be selected via the well-known integration formula for MISE to be minimized. In this section, we determine the TF domain expansion coefficients using the integration formula for optimal expansion of a function.

Let a square integrable time function $f(t)$ of Lebesgue measure be expanded into an m-term TF series in $t \in [0, T)$ as

$$f(t) \approx \sum_{i=0}^{m-1} a_i T1_i(t) + \sum_{i=0}^{m-1} b_i T2_i(t), \quad i = 0, 1, 2, \ldots, (m-1).$$

where, a_i's and b_i's are constant coefficients and $T1_i(t)$ and $T2_i(t)$ are given by equations (2.31) and (2.32).

If we consider an optimal representation of $f(t)$ leading to least integral squared error (ISE), the optimal coefficients a_i's and b_i's are to be determined conventionally from the following two equations:

$$\int_{ih}^{(i+1)h} f(t)T1_i(t)dt = a_i \int_{ih}^{(i+1)h} [T1_i(t)]^2 dt$$
$$+ b_i \int_{ih}^{(i+1)h} [T1_i(t)T2_i(t)]dt \quad (2.49)$$

$$\int_{ih}^{(i+1)h} f(t)T2_i(t)dt = a_i \int_{ih}^{(i+1)h} [T1_i(t)T2_i(t)]dt$$
$$+ b_i \int_{ih}^{(i+1)h} [T2_i(t)]^2 dt \quad (2.50)$$

Calling,

$$A \triangleq \int_{ih}^{(i+1)h} f(t)T1_i(t)dt \quad \text{and} \quad B \triangleq \int_{ih}^{(i+1)h} f(t)T2_i(t)dt$$

equations (2.49) and (2.50) become

$$A = a_i X_1 + b_i X_{12} \quad B = a_i X_{12} + b_i X_2 \quad (2.51)$$

where,

$$X_1 = \int_{ih}^{(i+1)h} [T1_i(t)]^2 dt$$
$$X_2 = \int_{ih}^{(i+1)h} [T2_i(t)]^2 dt$$
$$X_{12} = \int_{ih}^{(i+1)h} [T1_i(t)T2_i(t)]dt$$

Solving equation (2.51) we get

$$a_i = \frac{AX_2 - BX_{12}}{X_1 X_2 - X_{12}^2} \quad \text{and} \quad b_i = \frac{BX_i - AX_{12}}{X_1 X_2 - X_{12}^2} \quad (2.52)$$

Now we evaluate X_1, X_2, X_{12}, etc.

$$X_1 = \int_{ih}^{(i+1)h} [T1_i(t)]^2 dt = \int_{ih}^{(i+1)h} \left[1 - \left(\frac{t}{h} - i\right)\right]^2 dt = \frac{h}{3}$$

$$[\text{since, } T1_i(t) = (i+1) - \frac{t}{h}, \quad \text{for } ih \leq t < (i+1)h]$$

$$(2.53)$$

$$X_2 = \int_{ih}^{(i+1)h} [T2_i(t)]^2 dt = \int_{ih}^{(i+1)h} \left(\frac{t}{h} - i\right)^2 dt = \frac{h}{3} \quad (2.54)$$

and

$$X_{12} = \int_{ih}^{(i+1)h} [T1_i(t)][T2_i(t)]dt$$

$$= \int_{ih}^{(i+1)h} \left[1 - \left(\frac{t}{h} - i\right)\right]\left[\frac{t}{h} - i\right]dt = \frac{h}{6} \quad (2.55)$$

Putting these values of X_1, X_2, and X_{12} from equations (2.53) to (2.55) in equation (2.52) we have

$$\left.\begin{array}{l} a_i = \dfrac{A \cdot \frac{h}{3} - B \cdot \frac{h}{6}}{\frac{h^2}{9} - \frac{h^2}{36}} = \dfrac{2}{h}(2A - B) \\[4mm] b_i = \dfrac{B \cdot \frac{h}{3} - B \cdot \frac{h}{6}}{\frac{h^2}{12}} = \dfrac{2}{h}(2B - A) \end{array}\right\} \quad (2.56)$$

It is obvious that for piecewise linear functions, e.g., unit step function, ramp function etc., optimal or nonoptimal representation will become identical because of same a_i's and b_i's for the two cases and representational error (ISE) will be zero. For a unit step function, $a_i = b_i = 1$ for all i, and for a ramp function, $a_i = ih$ and $b_i = (i + 1)h$, for $i = 0, 1, 2, \ldots, (m - 1)$. In the next chapter, error estimates are computed and comparison of error for optimal and nonoptimal representation is presented.

2.9 Conclusion

The evolution and a few fundamental properties of the triangular orthogonal functions are discussed. The presented triangular functions use function samples as expansion coefficients, indicating a non-optimal approach. The optimal coefficients, that is, those determined by the traditional integration formula, have also been derived.

The set of triangular functions seems to be suitable for function approximation, approximate integration of functions and fit for application in the area of control system analysis and synthesis. These areas are explored in the following chapters.

References

1. Sansone, G., *Orthogonal functions*, Interscience, New York, 1959.
2. Chen, C. F. and Hsiao, C. H., A state space approach to Walsh series solution of linear systems, Int. J. Syst. Sci., vol. **6**, no. 9, pp. 833–858, 1975.
3. Chen, C. F. and Hsiao, C. H., Time domain synthesis via Walsh functions, IEE Proc., vol. **122**, no. 5, pp. 565–570, 1975.
4. Chen, C. F. and Hsiao, C. H., Design of piecewise constant gains for optimal control via Walsh functions, IEEE Trans. Automatic Control, vol. **AC-20**, no. 5, pp. 596–603, 1975.
5. Chen, C. F., Tsay, Y. T. and Wu, T. T., Walsh operational matrices for fractional calculus and their application to distributed systems, J. Franklin Instt., vol. **303**, no. 3, pp. 267–284, 1977.
6. Deb, A., Sarkar, G. and Sen, S. K., Block pulse functions, the most fundamental of all piecewise constant basis functions, Int. J. Syst. Sci., vol. **25**, no. 2, pp. 351–363, 1994.
7. Wang, Chi-Hsu, Generalized block pulse operational matrices and their applications to operational calculus, Int. J. Control, vol. **36**, no. 1, pp. 67–76, 1982.
8. Jiang, J. H. and Schaufelberger, W., Block pulse functions and their applications in control systems, LNCIS-179, Springer-Verlag, Berlin, 1992.

9. Deb, A., Sarkar, G., Bhattacharjee, M. and Sen, S. K., A new set of piecewise constant orthogonal functions for the analysis of linear SISO systems with sample-and-hold, J. Franklin Instt., vol. **335B**, no. 2, pp. 333–358, 1998.

10. Deb, A., Sarkar G. and Dasgupta, A., A complementary pair of orthogonal triangular function sets and its application to the analysis of SISO control systems, J. Instt. Engrs. (India), vol. **84**, December, pp. 120–129, 2003.

11. Tolstoy, G. P., Fourier series (English translation), Dover Publications, Inc., New York, 1976.

Chapter 3

Function Approximation via Triangular Function Sets and Operational Matrices in Triangular Function Domain

The proposed triangular function (TF) set may be utilized for approximating square integrable functions in a piecewise linear manner. The spirit of such approximation was explained in Section 2.5.

As was done with Walsh and block pulse functions, in this chapter, we use the complementary TF sets in a similar fashion for function approximation, and then form the operational matrices **P1** and **P2**, later used to solve differential as well as integral equations.

3.1 Approximation of a Square Integrable Time Function *f(t)* by BPF and TF

To compare the effectiveness of TF with that of BPF, let us consider the function

$$f(t) = \exp(-2t) \tag{3.1}$$

For $T = 1$ s and $m = 4$, the BPF representation of $f(t)$, using equations (2.20) and (2.21), is

$$f(t) \approx [0.7869387 \quad 0.4773024 \quad 0.2894986$$
$$0.1755898]\Psi_{(\mathbf{4})}(\mathbf{t}) \tag{3.2}$$

Similarly, using equations (2.34) and (2.35), triangular function representation of $f(t)$ is given by

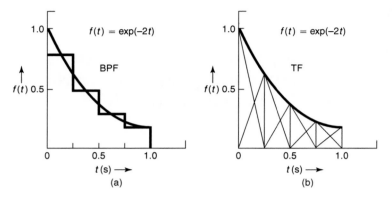

Figure 3.1. *(a) BPF representation and (b) TF representation of* $f(t) = \exp(-2t)$.

$$f(t) \approx [1 \quad 0.6065307 \quad 0.3678794 \quad 0.2231302]\mathbf{T1}_{(4)}(\mathbf{t})$$
$$+ [0.6065307 \quad 0.3678794 \quad 0.2231302$$
$$0.1353353]\mathbf{T2}_{(4)}(\mathbf{t}) \tag{3.3}$$

It is only apparent that TF representation is superior to BPF representation. Figure 3.1(a) and (b) compare $f(t)$ with its BPF and TF representations respectively.

3.2 Operational Matrices for Integration in Triangular Function Domain

Similar to the BPF domain technique, we now derive the operational matrix for integration in the orthogonal triangular function domain.

We integrate $T1_0$ of Fig. 2.6(a) and express the result in terms of an m-set triangular function using equations (2.34) and (2.35). That is, dropping the argument t for simplicity, we have

$$\int_0^t T1_0 d\tau = \int_0^t \left\{ \left(1 - \frac{\tau}{h}\right) u(\tau) + \frac{(\tau - h)}{h} u(\tau - h) \right\} d\tau$$
$$= \left(t - \frac{t^2}{2h}\right) u(t) - \left\{ \left(t - \frac{t^2}{2h}\right) - \frac{h}{2} \right\} u(t - h)$$

$$\approx \frac{h}{2}T2_0 + \frac{h}{2}[(T1_1 + T2_1) + (T1_2 + T2_2) + \cdots$$
$$+ \{T1_{(m-1)} + T2_{(m-1)}\}]$$
$$= \frac{h}{2}[0\ 1\ 1\ \ldots\ 1\ 1]\mathbf{T1}_{(\mathbf{m})}$$
$$+ \frac{h}{2}[1\ 1\ 1\ \ldots\ 1\ 1]\mathbf{T2}_{(\mathbf{m})} \tag{3.4}$$

It is obvious that integration of other members of the LHTF set will yield the same result, only the integrated curves will have different delays depending upon the respective delays of different component functions. Figure 3.2(a) shows the integration of the function $T1_0$.

Considering the above mentioned delays and using equation (3.4), we can write down the relation between the first integration of the LHTF set $\mathbf{T1}_{(\mathbf{m})}$ and the pair of triangular function sets $\mathbf{T1}_{(\mathbf{m})}$ and $\mathbf{T2}_{(\mathbf{m})}$ as

$$\int_0^t \mathbf{T1}_{(m)}d\tau = \frac{h}{2}[\![0\ 1\ 1\ 1\ \ldots\ \ldots\ 1\ 1]\!]_{(m \times m)}\mathbf{T1}_{(\mathbf{m})}$$
$$+ \frac{h}{2}[\![1\ 1\ 1\ 1\ \ldots\ \ldots\ 1\ 1]\!]_{(m \times m)}\mathbf{T2}_{(\mathbf{m})}$$
$$\overset{\triangle}{=} \mathbf{P1}_{(\mathbf{m})}\mathbf{T1}_{(\mathbf{m})} + \mathbf{P2}_{(\mathbf{m})}\mathbf{T2}_{(\mathbf{m})} \tag{3.5}$$

where we call the matrices $\mathbf{P1}_{(\mathbf{m})}$ and $\mathbf{P2}_{(\mathbf{m})}$ the operational matrices for integration in triangular function domain.

It is easy to see that, $\mathbf{P} = \mathbf{P1} + \mathbf{P2}$.

Now we integrate $T2_0$ to obtain the curve of Fig. 3.2(b).

Since integration is a linear operator, it is obvious that

$$\int_0^t T1_0 d\tau + \int_0^t T2_0 d\tau = \int_0^t (T1_0 + T2_0)d\tau = \int_0^t \Psi_0 d\tau \tag{3.6}$$

This is shown in Fig. 3.2 where the curve of Fig. 3.2(c) is obtained by adding the curves of Fig. 3.2(a) and (b).

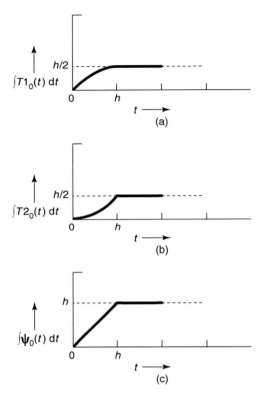

Figure 3.2. *(a) First integration of the triangular function $T1_0$, (b) First integration of the triangular function $T2_0$, (c) their relation to the first integration of BPF.*

Proceeding in a similar manner as for **T1**, we find that

$$\int_0^t \mathbf{T2}_{(m)}d\tau = \mathbf{P1}_{(m)}\mathbf{T1}_{(m)} + \mathbf{P2}_{(m)}\mathbf{T2}_{(m)} = \int_0^t \mathbf{T1}_{(m)}d\tau \tag{3.7}$$

Now, we attempt to integrate a square integrable function $f(t)$ in $t \in [0, T)$. Considering m component functions and a sampling period of h seconds, we can write, using equations (2.34) and (2.35)

$$f(t) \approx [f(0) \ f(h) \ f(2h) \ \ldots \ f((m-1)h)]\mathbf{T1}_{(m)}$$
$$+ [f(h) \ f(2h) \ f(3h) \ \ldots \ f(mh)]\mathbf{T2}_{(m)}$$
$$\triangleq \mathbf{C}^T\mathbf{T1}_{(m)} + \mathbf{D}^T\mathbf{T2}_{(m)} \tag{3.8}$$

Hence

$$\int_0^t f(\tau)d\tau \approx \mathbf{C}^T \int_0^t \mathbf{T1}_{(m)}d\tau + \mathbf{D}^T \int_0^t \mathbf{T2}_{(m)}dt$$
$$\approx (\mathbf{C} + \mathbf{D})^T\mathbf{P1T1}_{(m)} + (\mathbf{C} + \mathbf{D})^T\mathbf{P2T2}_{(m)} \tag{3.9}$$

where equations (3.5) and (3.7) are used.

Knowing $\mathbf{C}$ and $\mathbf{D}$, which are in fact $(m+1)$ number samples of $f(t)$ from 0 to T, we can determine approximate integration of $f(t)$.

3.2.1 Integration of a time function f(t) in BPF domain and TF domain

Let us consider the function $f(t) = \exp(-2t)$.

In BPF domain, approximate integration of $f(t)$, using equations (2.20), (2.21), and (2.23), is given by

$$\int f(t)dt \approx \mathbf{C}^T\mathbf{P}\mathbf{\Psi}_{(m)} \tag{3.10}$$

where, $\mathbf{C}, \mathbf{P}$, and $\mathbf{\Psi}_{(m)}$ are of order m.

Considering $T = 1$ s and $m = 4$, we have, from equations (2.20) to (2.23), the BPF domain representation of

$$\int_0^t f(t)dt$$

$$\approx [0.7869387 \ 0.4773024 \ 0.2894986 \ 0.1755898]\mathbf{P}_{(4)}\mathbf{\Psi}_{(4)}$$
$$= [0.0983673 \ 0.2563975 \ 0.3522476 \ 0.4103837]\mathbf{\Psi}_{(4)} \tag{3.11}$$

Considering $T = 1$ s and $m = 4$, TF domain representation of $f(t)$, using equations (2.34) and (2.35), is

$$f(t) \approx [1 \ 0.6065307 \ 0.3678794 \ 0.2231302]\mathbf{T1}_{(4)}$$
$$+ [0.6065307 \ 0.3678794 \ 0.2231302 \ 0.1353353]\mathbf{T2}_{(4)}$$
$$\triangleq \mathbf{C1}^T\mathbf{T1}_{(4)} + \mathbf{D1}^T\mathbf{T2}_{(4)} \tag{3.12}$$

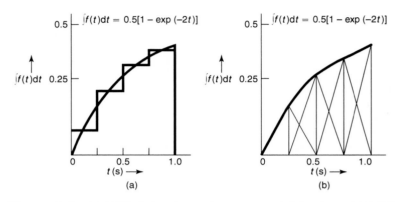

Figure 3.3. *(a) BPF domain representation of ∫ f(t)dt and (b) TF domain representation of ∫ f(t)dt.*

Table 3.1. Comparison of BPF and TF domain integration for the function $f(t) = \exp(-2t)$

Times t (s)	Samples of $\int_0^t \exp(-2t)dt$	BPF domain solution	TF domain solution	TF domain % error
0	0		0	–
		0.0983673		
0.25	0.1967347		0.2008163	−2.07452
		0.2563975		
0.5	0.3160603		0.3226176	−2.07470
		0.3522476		
0.75	0.3884349		0.3964938	−2.07471
		0.4103837		
1.0	0.4323324		0.4413020	−2.07470

Using equations (3.9)

$$\int_0^t f(t)dt \approx [0 \quad 0.2008163 \quad 0.3226176 \quad 0.3964938]\mathbf{T1}_{(4)}$$

$$+ [0.2008163 \quad 0.3223176 \quad 0.3964938$$
$$0.4413020]\mathbf{T2}_{(4)} \qquad (3.13)$$

Figure 3.3 shows BPF and TF domain representations of ∫ f(t)dt, while Table 3.1 compares the efficiency of integration in BPF and TF domain and computes percentage error in the TF domain.

3.3 Error Analysis

The representational error for equal width block pulse function expansion of any square integrable function of Lebesgue measure has been investigated by Rao and Srinivasan [1], while the error analysis for pulse-width modulated generalized block pulse function (PWM-GBPF) expansion has been carried out by Deb et al. [2] To present an error analysis in the triangular function domain, we follow a similar track.

Let us consider $(m + 1)$ sample points of the function $f(t)$, having a sampling period h, denoted by $f(ih), i = 0, 1, 2, \ldots, m$. Then piecewise linear representation of the function $f(t)$ by triangular functions is obtained simply by joining these sample points. The equation of one such straight line in the ith interval is given by

$$\hat{f}(t) = m_i t + f(ih) - i m_i h \qquad (3.14)$$

where,

$$m_i = \frac{f[(i + 1)h] - f(ih)}{h} \qquad (3.15)$$

Then integral squared error (ISE) in the ith interval is given by

$$[E_i]^2 \triangleq \int_{ih}^{(i+1)h} \left[\hat{f}(t) - f(t) \right]^2 dt \qquad (3.16)$$

Let the function $f(t)$ be expanded by Taylor series in the ith interval around the point μ_i considering second-order approximation

$$f(t) \approx f(\mu_i) + \dot{f}(\mu_i)(t - \mu_i) + \ddot{f}(\mu_i)(t - \mu_i)^2 / 2! \qquad (3.17)$$

where, $\mu_i \in [ih, (i + 1)h]$.

Equation (3.16) may be written as

$$[E_i]^2 = \int_{ih}^{(i+1)h} \left[\{ f(ih) - i m_i h - f(\mu_i) + \mu_i \dot{f}(\mu_i) \} \right.$$
$$\left. + \{ m_i - \dot{f}(\mu_i) \} t - \ddot{f}(\mu_i)(t - \mu_i)^2 / 2! \right]^2 dt$$
$$= \int_{ih}^{(i+h)h} \left[A + Bt + C(t - \mu_i)^2 \right]^2 dt \qquad (3.18)$$

where,

$$A \triangleq f(ih) - im_i h - f(\mu_i) + \mu_i \dot{f}(\mu_i)$$
$$B \triangleq m_i - \dot{f}(\mu_i)$$
$$C \triangleq -\ddot{f}(\mu_i)/2! \qquad (3.19)$$

Hence, equation (3.18) may be simplified to

$$
\begin{aligned}
[E_i]^2 = {} & A^2 h + \frac{B^2 h^3}{3}(3i^2 + 3i + 1) \\
& + \frac{C^2 h^5}{5}(5i^4 + 10i^3 + 10i^2 + 5i + 1) \\
& - h^4 \mu_i C^2 (4i^3 + 6i^2 + 4i + 1) + 2h^3 \mu_i^2 C^2 (3i^2 + 3i + 1) \\
& - 2h^2 \mu_i^3 C^2 (2i + 1) + hC^2 \mu_i^4 + ABh^2(2i + 1) \\
& + \frac{h^4 BC}{2}(4i^3 + 6i^2 + 4i + 1) - \frac{4h^3 BC\mu_i}{3}(3i^2 + 3i + 1) \\
& + h^2 BC\mu_i^2(2i + 1) + \frac{2h^3 CA}{3}(3i^2 + 3i + 1) \\
& - 2h^2 AC\mu_i(2i + 1) + 2hAC\mu_i^2 \qquad (3.20)
\end{aligned}
$$

Then the upper bound of ISE over m subintervals is given by

$$
\begin{aligned}
\sum_{i=0}^{m-1} [E_i]^2_{\max} \triangleq E^2 = {} & \ddot{f}^2_{\max} \left[\frac{m^5 h^5}{20} - \frac{m^4 h^4}{4}\mu_{\max} + \frac{m^3 h^3}{2}\mu^2_{\max} \right. \\
& \left. - \frac{m^2 h^2}{2}\mu^3_{\max} + \frac{mh}{4}\mu^4_{\max} \right] \\
& + \dot{f}^2_{\max} \left[\frac{(2m^3 - 3m^2 + m)h^3}{6} \right. \\
& \left. - (m^2 - m)h^2 \mu_{\max} + mh\mu^2_{\max} \right]
\end{aligned}
$$

$$+ \ddot{f}_{max}\dot{f}_{max}\left[\frac{(3m^4 - 2m^3 - m^2)h^4}{12} \right.$$

$$- \frac{(6m^3 - 3m^2 - m)h^3}{6}\mu_{max}$$

$$\left. + \frac{(3m^2 - m)h^2}{2}\mu_{max}^2 - mh\mu_{max}^3 \right] \quad (3.21)$$

where

$$f_{max} \triangleq \max\{f(ih), f[(i + 1)h], f(\mu_i)\}$$

$$\dot{f}_{max} \triangleq \max\{\dot{f}(\mu_i), m_i\}$$

$$\ddot{f}_{max} \triangleq \max\{\ddot{f}(\mu_i)\}$$

and $\quad \mu_{max} \triangleq \max\{\mu_i\}$

These maximum values are considered to be the largest over the entire period. So, equation (3.21) gives the upper bound of ISE.

Let us now consider some special cases that can be derived from equation (3.20).

For simplicity, assume that the function is approximated by first-order Taylor series expansion. In that case, $\ddot{f}(\mu_i) = 0$. Then from equation (3.19), $C = 0$.

Equation (3.20) now can be simplified to

$$[E_i]^2 = A^2h + \frac{B^2h^3}{3}(3i^2 + 3i + 1) + ABh^2(2i + 1) \quad (3.22)$$

Case I: Let us assume that the function $f(t)$ is a step function. Then, $f(ih) = f(\mu_i), m_i = 0, \dot{f}(\mu_i) = 0$. This implies, from equation (3.19), $A = 0$ and $B = 0$. Hence, from equation (3.22), we have ISE in the ith inteval, i.e., $[E_i]^2 = 0$.

Case II: If $f(t)$ is a ramp function, then $m_i = m_r$(constant). Since $m_r = \dot{f}(\mu_i)$, we have $B = 0$. Consider $\mu_i = ih + x, 0 \le x \le h$. Then

$$f(\mu_i) = f[(ih) + x] = f(ih) + m_r x$$

Hence, from equation (3.19), we get

$$A = f(ih) - im_r h - f(ih) - m_r x + [ih + x]m_r = 0$$

Then, from (3.22), ISE in the ith interval is zero.

Case III: If $f(t)$ is a piecewise ramp function having a slope m_i in the ith interval, then we have

$$m_i = \dot{f}(\mu_i), \quad \text{and} \quad \mu_i = ih + x, \quad 0 \le x \le h.$$

It follows from equation (3.19) that $A = B = 0$, implying zero ISE.

As the result of **cases II** and **III** are independent of x, we can conclude that for a ramp function—whether continuous or piecewise—ISE is zero irrespective of the magnitude of μ_i, the focal point of Taylor series expansion.

3.4 Comparison of Error for Optimal and Nonoptimal Representation via Block Pulse as well as Triangular Functions

To compare the error for optimal and nonoptimal triangular function representation of a square integrable time function, we take help of Sections 2.8 and 3.1.

Let $f(t) = \exp(-t)$

Then from equations (2.49) and (2.50), putting the expressions for $T1_i(t)$ and $T2_i(t)$, we can write

$$
\begin{aligned}
A &= \int_{ih}^{(i+1)h} f(t)T1_i(t)dt \\
&= \int_{ih}^{(i+1)h} e^{-t}\left[(1+i) - \frac{t}{h}\right]dt = \frac{e^{-ih}}{h}[-1 + h + e^{-h}]
\end{aligned}
\tag{3.23}
$$

and

$$
\begin{aligned}
B &= \int_{ih}^{(i+1)h} f(t)T2_i(t)dt \\
&= \int_{ih}^{(i+1)h} e^{-t}\left[\frac{t}{h} - i\right]dt = \frac{e^{-ih}}{h}[-e^{-h} - he^{-h} + 1]
\end{aligned}
\tag{3.24}
$$

Table 3.2. Comparison of TF domain optimal and nonoptimal coefficients for the function $f_1(t) = \exp(-t)(T = 4\,\text{s}, h = 1\,\text{s},$ and $m = 4)$

	Optimal coefficients		Nonoptimal coefficients	
Time t (s)	a_i	b_i	a_i'	b_i'
0	0.9430	0.3212	1.0000	0.3679
1	0.3469	0.1182	0.3679	0.1353
2	0.1276	0.0435	0.1353	0.0498
3	0.04695	0.0160	0.0498	0.0183

Putting these values in equation (2.56) and simplifying

$$a_i = \left(\frac{2}{h^2}\right) \exp(-ih)(-3 + 2h + 3e^{-h} + he^{-h}) \qquad (3.25)$$

$$b_i = \left(\frac{2}{h^2}\right) \exp(-ih)\{-3\exp(-h) + 3 - 2h\exp(-h) - h\}$$

$$(3.26)$$

3.4.1 Examples

Let us consider $T = 4\,\text{s}$, $h = 1\,\text{s}$, and $m = 4$.

Substituting the value of h in the above expressions, we get four a_i's and b_i's as tabulated in Table 3.2. The other two columns of Table 3.2 show the sample values of the function $\exp(-t)$ to show the deviation between optimal and nonoptimal one. Similarly, replacing function $f(t)$ by t^2 instead of $\exp(-t)$, Table 3.3 can be formed.

Calling the optimal coefficients a_i, b_i, and nonoptimal coefficients a_i', b_i', in Tables 3.2 and 3.3, the values of respective coefficients for two functions $f_1(t) = \exp(-t)$ and $f_2(t) = t^2$ are compared.

3.4.2 Integral squared error (ISE)

Let us consider $T = 4$ s, $h = 1$ s, and $m = 4$.

The integral squared error (ISE) for the functions $f_1(t) = \exp(-t)$ and $f_2(t) = t^2$ are estimated using equation (3.16) for

Table 3.3. Comparison of TF domain optimal and nonoptimal coefficients for the function $f_2(t) = t^2$ ($T = 4$ s, $h = 1$ s, and $m = 4$)

Time t (s)	Optimal coefficients		Nonoptimal coefficients	
	a_i	b_i	a_i'	b_i'
0	−0.1667	0.8333	0.0000	1.0000
1	0.8333	3.8333	1.0000	4.0000
2	3.8333	8.8333	4.0000	9.0000
3	8.8333	15.8333	9.0000	16.0000

Table 3.4. Representational error for four different functions with different choices for basis functions ($T = 4$ s, $h = 1$ s, and $m = 4$)

Basis functions	Representational error (ISE) for			
	$\exp(-t)$	t^2	t	$u(t)$
TF(optimal)	6.1657×10^{-4}	2.2235×10^{-2}	0	0
TF(nonoptimal)	3.7318×10^{-3}	0.1334	0	0
BPF	3.7870×10^{-2}	7.0222	0.3333	0
SHF	0.1943	25.4666	1.3333	0

both optimal and nonoptimal representations via BPF, sample-and-hold functions (SHF) and TF over the time interval specified above. As indicated in *section 2.5*, BPF and SHF function representations of any square integrable function $f(t)$ also lead to optimal and non-optimal results, respectively.

Using equation (3.21), ISEs for representation via optimal and nonoptimal triangular functions may be determined for four typical functions, namely, $u(t)$, t, $\exp(-t)$, and t^2. Also MISEs for representation of the same functions via BPF [1] and SHF [3] domains are computed. All ISEs are tabulated and compared in Table 3.4. As expected, ISE is the least for TF (optimal) representation and the worst for SHF approximation. For a unit step function, the representational error is zero for all choices because the unit step function is piecewise linear and piecewise constant as well.

3.5 Conclusion

The pair of orthogonal triangular function (TF) sets have been employed for piecewise linear approximations of time functions of Lebesgue measure. Further, operational matrices for integration in the TF domain, namely **P1** and **P2**, have been derived and their relation with the well-known block pulse function domain integral operational matrix **P** is shown. It has also been established with illustration that integration of time functions with **P1** and **P2** are more accurate than that in the BPF domain.

A detailed study of the representational error has also been made to estimate the upper bound of the ISE for the TF approximation of a time function $f(t)$ of Lebesgue measure.

The presented analysis is based upon nonoptimal determination of the TF expansion coefficients to enjoy the advantages of using function samples as the coefficients. Comparison of optimal and nonoptimal expansion coefficients with related ISE of a few standard functions—using triangular functions, block pulse functions and sample-and-hold functions—are also presented. It is observed that though ISE is larger for non-optimal TF domain analysis, advantages of using function samples outweigh the slight loss in accuracy. For the case of BPF versus SHF, it is noted that use of SHF for sample-and-hold systems is always a better bargain than using the BPF set. However, with decreasing h, the nonoptimal BPF domain solution tends to coincide with the optimal BPF domain solution.

References

1. Rao, G. P. and Srinivasan, T., Analysis and synthesis of dynamic systems containing time delays via block pulse functions, Proc. IEE, vol. **125**, no. 9, pp. 1064–1068, 1978.
2. Deb, A., Sarkar, G. and Sen, S. K., Linearly pulse-width modulated block pulse functions and their application to linear SISO feedback control system identification, IEE Proc. Control Theory and Appl., vol. **142**, no. 1, pp. 44–50, 1995.
3. Deb, A., Sarkar, G., Bhattacharjee, M. and Sen, S. K., A new set of piecewise constant orthogonal functions for the analysis of linear SISO systems with sample-and-hold, J. Franklin Instt., vol. **335B**, no. 2, pp. 333–358, 1998.

Chapter 4

Analysis of Dynamic Systems via State Space Approach

Modern control theory [1,2] is contrasted with conventional control theory in that the former is applicable to multi-input multi-output (MIMO) systems, which may be linear or nonlinear, time invariant or time varying, while the latter is applicable only to linear time invariant single-input single-output (SISO) systems. Also, modern control theory is essentially a time domain approach, while conventional control theory is a complex frequency domain approach.

The state of a dynamic system is the smallest set of variables such that the knowledge of these variables at $t = t_0$, together with the input for $t \geq t_0$, completely determines the behavior of the system for any time $t \geq t_0$.

A dynamic system consisting of a finite number of lumped elements may be described by ordinary differential equations in which time is the independent variable. By use of vector-matrix notation, an nth order differential equation may be expressed by a first-order vector-matrix differential equation. If n elements of the vector are a set of state variables, then the vector-matrix differential equation is called state equation.

A linear time-invariant SISO control system may be modeled by

$$\dot{\mathbf{x}}(\mathbf{t}) = \mathbf{Ax}(\mathbf{t}) + \mathbf{Bu}(\mathbf{t}) \tag{4.1}$$

$$\mathbf{y}(\mathbf{t}) = \mathbf{Cx}(\mathbf{t}) + \mathbf{Du}(\mathbf{t}) \tag{4.2}$$

where the first-order differential equation (4.1) is the state equation and the algebraic equation (4.2) is the output equation, the variables having usual significances.

Block pulse functions [3,4] were introduced by Chen et al. [5] and have been extensively used for the analysis [6,7] and identification [6–9]. New function sets [10,11] were also employed for such task and some of the methods were not attractive computationally.

We now apply the newly proposed piecewise linear triangular function sets (both LHTF and RHTF) [12] in the analysis of dynamic systems via state space approach.

4.1 Analysis of Dynamic Systems via Triangular Functions

Consider the time-invariant linear SISO dynamic system modeled by [2]

$$\dot{\mathbf{x}} = \mathbf{A}\mathbf{x} + \mathbf{B}\mathbf{u} \quad \text{and} \quad \mathbf{x(0)} = \mathbf{x_0} \tag{4.3}$$

where, $\mathbf{x}$ is an n-component state vector, $\mathbf{u}$ is an r-component input vector, and $\mathbf{A}$ and $\mathbf{B}$ are matrices of appropriate dimensions. To find out the block pulse domain solution [6], the rate vector $\dot{\mathbf{x}}$, the state vector $\mathbf{x}$ and the overall control vector $\mathbf{Bu}$ are expanded into a set of m-term block pulse functions over the semi-open interval $[0, T)$ as

$$\left.\begin{array}{l} \dot{\mathbf{x}}(\mathbf{t}) = \displaystyle\sum_{i=0}^{m-1} \mathbf{C}_{i+1}^{\mathsf{T}} \mathbf{\Psi}_i(t) \triangleq \mathbf{C}\mathbf{\Psi}(\mathbf{t}) \\[4mm] \mathbf{x}(\mathbf{t}) = \displaystyle\sum_{i=0}^{m-1} \mathbf{D}_{i+1}^{\mathsf{T}} \mathbf{\Psi}_i(t) \triangleq \mathbf{D}\mathbf{\Psi}(\mathbf{t}) \\[4mm] \mathbf{Bu}(\mathbf{t}) = \displaystyle\sum_{i=0}^{m-1} \mathbf{E}_{i+1}^{\mathsf{T}} \mathbf{\Psi}_i(t) \triangleq \mathbf{E}\mathbf{\Psi}(\mathbf{t}) \end{array}\right\} \tag{4.4}$$

Here, $\mathbf{C_{i+1}}, \mathbf{D_{i+1}}$, and $\mathbf{E_{i+1}}$ are n-vectors and form the $(i+1)$th column of the $n \times m$ matrices $\mathbf{C}, \mathbf{D}$, and $\mathbf{E}$, respectively. For a given input $u(t)$, $\mathbf{E}$ is known by using equation (2.21). To solve the state vector $\mathbf{x}$ in the BPF domain, it is required to solve for the matrix $\mathbf{D}$ only. The following identity is used for obtaining the solution

$$\int_0^t \dot{\mathbf{x}}(\tau)\mathbf{d}\tau = \mathbf{x}(t) - \mathbf{x}(0) \tag{4.5}$$

Using BPF domain operational matrix $\mathbf{P}$ in equation (4.5), we have

$$\mathbf{CP\Psi}(t) = \mathbf{D\Psi}(t) - \mathbf{x}(0) = \mathbf{D\Psi}(t) - \tilde{\mathbf{x}}_0\mathbf{\Psi}(t) \tag{4.6}$$

where, $\tilde{\mathbf{x}}_0$ is an $n \times m$ matrix each column of which is $\mathbf{x}_0$.

Also, from equations (4.3) and (4.4), we have

$$\mathbf{C\Psi}(t) = \mathbf{AD\Psi}(t) + \mathbf{E\Psi}(t) \tag{4.7}$$

Equating like coefficients of $\Psi_i(t)$ in equations (4.6) and (4.7), the $\mathbf{D}$ vector is solved as [6]

$$\left. \begin{aligned} \mathbf{D}_1 &= \left[\mathbf{I} - \tfrac{\mathbf{A}T}{2m}\right]^{-1}\mathbf{x}_0 + \tfrac{T}{2m}\left[\mathbf{I} - \tfrac{\mathbf{A}T}{2m}\right]^{-1}\mathbf{E}_1 \\ \mathbf{D}_{i+2} &= \left[\mathbf{I} - \tfrac{\mathbf{A}T}{2m}\right]^{-1}\left[\mathbf{I} + \tfrac{\mathbf{A}T}{2m}\right]\mathbf{D}_{i+1} + \tfrac{T}{2m}\left[\mathbf{I} - \tfrac{\mathbf{A}T}{2m}\right]^{-1} \\ &\quad \times [\mathbf{E}_{i+1} + \mathbf{E}_{i+2}] \qquad\qquad \text{for all } i \geq 0 \end{aligned} \right\} \tag{4.8}$$

Solution of the difference equation (4.8) provides an m-term approximation of $\mathbf{x}(t)$ in the BPF domain.

Proceeding in a similar manner, we try to solve for $\mathbf{x}(t)$ in the triangular function domain.

Following equation (2.34), the rate vector $\dot{\mathbf{x}}$, the state vector $\mathbf{x}$ and the overall control vector $\mathbf{Bu}$ are expanded into triangular function domain as given below:

$$\left. \begin{aligned} \dot{\mathbf{x}}(t) &= \sum_{i=0}^{m-1} \mathbf{C1}_{i+1}^{\mathrm{T}}T1_i(t) + \sum_{i=0}^{m-1} \mathbf{C2}_{i+1}^{\mathrm{T}}T2_i(t) \triangleq \mathbf{C1T1}(t) + \mathbf{C2T2}(t) \\ \mathbf{x}(t) &= \sum_{i=0}^{m-1} \mathbf{D1}_{i+1}^{\mathrm{T}}T1_i(t) + \sum_{i=0}^{m-1} \mathbf{D2}_{i+1}^{\mathrm{T}}T2_i(t) \triangleq \mathbf{D1T1}(t) + \mathbf{D2T2}(t) \\ \mathbf{Bu}(t) &= \sum_{i=0}^{m-1} \mathbf{E1}_{i+1}^{\mathrm{T}}T1_i(t) + \sum_{i=0}^{m-1} \mathbf{E2}_{i+1}^{\mathrm{T}}T2_i(t) \triangleq \mathbf{E1T1}(t) + \mathbf{E2T2}(t) \end{aligned} \right\}$$
$$\tag{4.9}$$

Here, $\mathbf{C1}_{i+1}, \mathbf{C2}_{i+1}, \mathbf{D1}_{i+1}, \mathbf{D2}_{i+1}, \mathbf{E1}_{i+1}$ and $\mathbf{E2}_{i+1}$ are n-vectors and form the $(i + 1)$th column of the $n \times m$ matrices

C1, **C2**, **D1**, **D2**, **E1** and **E2** respectively. For a given input **u**(t), **El** and **E2** are known using equation (2.35). To solve the state vector **x** in the **TF** domain, equation (4.5) is used.

Using equation (4.9) in equation (4.5) and the property of equation (3.7), we get

$$[CI + C2] \int T1(t)dt = D1T1(t) + D2T2(t) - \tilde{x}_0 \quad (4.10)$$

Using equations (3.7) and (4.3) in equation (4.10), and dropping the argument (t), we can write

$$[CI + C2][P1T1 + P2T2] = D1T1 + D2T2 - \tilde{x}_0[T1 + T2]$$

Equating the like coefficients of the basis functions **T1** and **T2**, we obtain

$$[C1 + C2]P1 = D1 - \tilde{x}_0 \quad (4.11)$$
$$[C1 + C2]P2 = D2 - \tilde{x}_0 \quad (4.12)$$

Adding equations (4.11) and (4.12), we get

$$[C1 + C2][P1 + P2] = [D1 + D2] - 2\tilde{x}_0$$

Putting, $C1 + C2 \overset{\Delta}{=} C$, $D1 + D2 \overset{\Delta}{=} D$ and using the relation $P1 + P2 = P$, we have

$$CP = D - 2\tilde{x}_0 \quad (4.13)$$

Equating the $(i + 1)$th column, we get

$$(CP)_{i+1} = D_{i+1} - 2x_0 \quad (4.14)$$

Again by putting equation (4.9) in equation (4.3), we can write

$$C1T1 + C2T2 = A[D1T1 + D2T2] + [E1T1 + E2T2]$$

Equating the like coefficients of the basis functions **T1** and **T2**, we obtain

$$C1 = AD1 + E1 \quad \text{and} \quad C2 = AD2 + E2$$

Adding, we get

$$\mathbf{C} = \mathbf{AD} + \mathbf{E} \qquad (4.15)$$

Equating the $(i+1)$th column, we have

$$\mathbf{C}_{i+1} = (\mathbf{AD})_{i+1} + \mathbf{E}_{i+1} \qquad (4.16)$$

It is noted that equations (4.13) and (4.15) are similar to equations (4.6) and (4.7).

Had we considered the vectors **C**, **D**, and **E** to be defined as follows

$$\mathbf{C} = \frac{1}{2}[\mathbf{C1} + \mathbf{C2}], \quad \mathbf{D} = \frac{1}{2}[\mathbf{D1} + \mathbf{D2}] \text{ and } \mathbf{E} = \frac{1}{2}[\mathbf{E1} + \mathbf{E2}]$$
$$(4.17)$$

then equations (4.13) and (4.15) would have been identical to equations (4.6) and (4.7). This may induce one to think that triangular function solution may be available directly from block pulse functions, knowing initial values of respective functions. However, this would not lead to the BPF solution as derived in Ref. [6]. A little consideration reveals that the solution obtained using relations (4.17) is a BPF domain solution in a nonoptimal sense, where the coefficents are derived from the average values of the samples of concerned time functions. Thus, these vectors **C**, **D** and **E** of equations (4.17) are different from those of equations (4.6) and (4.7). While the vectors of equations (4.6) and (4.7) are optimal in the least mean square error sense, vectors of equation (4.7) are not. In fact, there are innumerable ways of forming a nonoptimal BPF set (like above) of which another example is the sample and hold functions [11].

A nonoptimal BPF domain solution (let us call this specific approach BPFN) using equations (4.17) does not offer a piecewise linear solution like the TF domain solution. Also, the integral squared error (ISE) with such nonoptimal set is much larger than that incurred with TF approach (vide Section 3.4).

To proceed with the TF domain solution, we use the property of the matrix $\mathbf{P}$ and can write

$$(\mathbf{CP})_{i+1} = \frac{T}{m} \sum_{j=1}^{i} \mathbf{C}_j + \frac{T}{2m} \mathbf{C}_{i+1}, \quad i \geq j$$

Writing in a recurrence relation form

$$(\mathbf{CP})_1 = \frac{T}{2m} \mathbf{C}_1$$

$$(\mathbf{CP})_{i+2} = (\mathbf{CP})_{i+1} + \frac{T}{2m}(\mathbf{C}_{i+1} + \mathbf{C}_{i+2}) \quad \text{for all } i \geq 0 \quad (4.18)$$

Combining equations (4.14) and (4.18), we get solutions for the state vector as

$$\mathbf{D}_1 = \frac{T}{2m} \mathbf{C}_1 + 2\mathbf{x_0}$$

$$\mathbf{D}_{i+2} = \mathbf{D}_{i+1} + \frac{T}{2m}(\mathbf{C}_{i+1} + \mathbf{C}_{i+2}) \quad \text{for all } i \geq 0 \quad (4.19)$$

Substituting $\mathbf{C_{i+1}}$ from equation (4.16) in equations (4.17), we have the solution vector $\mathbf{D}$ as

$$\left. \begin{array}{l} \mathbf{D}_1 = 2\left[\mathbf{I} - \frac{\mathbf{A}T}{2m}\right]^{-1} \mathbf{x_0} + \frac{T}{2m}\left[\mathbf{I} - \frac{\mathbf{A}T}{2m}\right]^{-1} \mathbf{E}_1 \\[2mm] \mathbf{D}_{i+2} = \left[\mathbf{I} - \frac{\mathbf{A}T}{2m}\right]^{-1}\left[\mathbf{I} + \frac{\mathbf{A}T}{2m}\right]\mathbf{D}_{i+1} \\[2mm] \qquad + \frac{T}{2m}\left[\mathbf{I} - \frac{\mathbf{A}T}{2m}\right]^{-1}[\mathbf{E}_{i+1} + \mathbf{E}_{i+2}] \quad \text{for all } i \geq 0 \end{array} \right\} \quad (4.20)$$

Solution of the difference equation (4.20) provides an m-term approximation of $\mathbf{x}(t)$ in the TF domain.

Comparing equations (4.8) and (4.20), we find that both the solutions are exactly identical except the first term of $\mathbf{D}_1$.

So, general expression for $\mathbf{D}_1$ can be written as

$$\mathbf{D}_1 = K\left[\mathbf{I} - \frac{\mathbf{A}T}{2m}\right]^{-1} \mathbf{x_0} + \frac{T}{2m}\left[\mathbf{I} - \frac{\mathbf{A}T}{2m}\right]^{-1} \mathbf{E}_1 \quad (4.21)$$

where, $K = 1$ for nonoptimal BPF domain solution, and $K = 2$ for TF domain solution.

4.2 Numerical Experiment [2]

Consider the time invariant linear SISO system described by equation (4.3), where

$$\mathbf{A} = \begin{bmatrix} 0 & 1 \\ -2 & -3 \end{bmatrix}, \mathbf{B} = \begin{bmatrix} 0 \\ 1 \end{bmatrix}, \quad \text{and } \mathbf{x}(0) = 0$$

The actual solution of the states is

$$x_1(t) = 0.5 - \exp(-t) + 0.5\exp(-2t)$$

and $$x_2(t) = \exp(-t) - \exp(-2t)$$

For a unit step input occurring at $t = 0$, using equations (2.20) and (2.21), the matrices **E1** and **E2** are given by

$$\mathbf{E1} = \mathbf{E2} = \begin{bmatrix} 0 & 0 & 0 & 0 & 0 & 0 & 0 & 0 & 0 & 0 \\ 1 & 1 & 1 & 1 & 1 & 1 & 1 & 1 & 1 & 1 \end{bmatrix}$$

where we have considered $m = 10$, $T = 1$s, and $h = 0.1$ s.

Using the recurrence relation of equation (4.20) we have

$$\mathbf{D}_{\text{TF}} = \begin{bmatrix} 0.0043290 & 0.0204456 & 0.0493381 & 0.0871877 \\ 0.0865801 & 0.2357527 & 0.3420969 & 0.4148955 \end{bmatrix}$$

$$\begin{bmatrix} 0.1310126 & 0.1785017 & 0.2278811 & 0.2778047 \\ 0.4616010 & 0.4881821 & 0.4994056 & 0.4990663 \end{bmatrix}$$

$$\begin{bmatrix} 0.3272667 & 0.3755304 \\ 0.4901732 & 0.4751022 \end{bmatrix}$$

As the first column of **D1** contains the initial values of the states $x_1(0)$ and $x_2(0)$, it is easy to decouple **D1** and **D2** from **D**. That is

$$\mathbf{D1} = \begin{bmatrix} 0.0000000 & 0.0043290 & 0.0161166 & 0.0332215 \\ 0.0000000 & 0.0865801 & 0.1491726 & 0.1929243 \end{bmatrix}$$

$$\begin{bmatrix} 0.0539663 & 0.0770463 & 0.1014554 & 0.1264257 \\ 0.2219712 & 0.2396298 & 0.2485523 & 0.2508533 \end{bmatrix}$$

$$\begin{bmatrix} 0.1513790 & 0.1758877 \\ 0.2482129 & 0.2419603 \end{bmatrix}$$

$$\mathbf{D2} = \begin{bmatrix} 0.0043290 & 0.0161166 & 0.0332215 & 0.0539663 \\ 0.0865801 & 0.1491726 & 0.1929243 & 0.2219712 \end{bmatrix}$$

$$0.0770463 \quad 0.1014554 \quad 0.1264257 \quad 0.1513790$$
$$0.2396298 \quad 0.2485523 \quad 0.2508533 \quad 0.2482129$$

$$\begin{matrix} 0.1758877 & 0.1996428 \\ 0.2419603 & 0.2331419 \end{matrix} \tag{4.22}$$

The triangular function domain solution is compared with time samples of $x_1(t)$ and $x_2(t)$ and found to be reasonably close. The ISE for $x_1(t)$ and $x_2(t)$ were found to be 1.096349E-07 and 9.099104E-07, respectively.

For $m = 10, T = 1$ s and $h = 0.1$ s, the nonoptimal BPF domain solution for the same system is

$$\mathbf{D}_{\mathrm{BPFN}} = \begin{bmatrix} 0.0021645 & 0.0102228 & 0.0246691 & 0.0435939 \\ 0.0432900 & 0.1178764 & 0.1710484 & 0.2074477 \end{bmatrix}$$

$$0.0655063 \quad 0.0892509 \quad 0.1139406 \quad 0.1389023$$
$$0.2308005 \quad 0.2440910 \quad 0.2497028 \quad 0.2495331$$

$$\begin{matrix} 0.1636333 & 0.1877652 \\ 0.2450866 & 0.2375511 \end{matrix}$$

In this case, the ISEs for $x_1(t)$ and $x_2(t)$ were found to be 3.695390E-05 and 1.218721E-04, respectively.

From these facts we have, for $x_1(t)$

$$\frac{\mathrm{ISE}_{\mathrm{BPFN}}}{\mathrm{ISE}_{\mathrm{TF}}} = 337.0632697 \tag{4.23}$$

and for $x_2(t)$

$$\frac{\mathrm{ISE}_{\mathrm{BPFN}}}{\mathrm{ISE}_{\mathrm{TF}}} = 133.9385994 \tag{4.24}$$

From the ratios of ISEs, it is easy to note that TF domain solution is superior to nonoptimal BPF domain solution. Also, the lesser magnitude of the ratio in the second case indicates that $x_2(t)$ is a relatively smoother function than $x_1(t)$. This is quite evident from the analytical expressions of the time functions.

4.3 Conclusion

With the help of **P1** and **P2**, the triangular function sets have been applied to the analysis of linear SISO dynamic systems, and the estimated ISE is found to be much less than the alternative staircase solution obtained from BPF domain analysis.

A detailed study of the representational error has been presented in Section 3.3.

All computer programs developed for this work are in double precision mode, and as indicated in equation (4.21), a general computer program was developed with "choice" facility depending on the value of K to analyze the given system in either nonoptimal BPF domain, or in TF domain. However, with increasing m, the nonoptimal BPF domain solution will tend to coincide with the optimal BPF domain solution.

References

1. Ogata, K., State space analysis of control systems, Prentice-Hall Inc., Englewood Cliffs, NJ, 1967.
2. Ogata, K., *Modern control engineering* (4th ed.), Pearson Education Asia, New Delhi, 2002.
3. Jiang, J. H. and Schaufelberger, W., *Block pulse functions and their applications in control systems*, LNCIS-179, Springer-Verlag, Berlin, 1992.
4. Deb, A., Sarkar, G. and Sen, S. K., Block pulse functions, the most fundamental of all piecewise constant basis functions, Int. J. Syst. Sci., vol. **25** no. 2. pp. 351–363, 1994.
5. Chen, C. F., Tsay, Y. T. and Wu, T. T., Walsh operational matrices for fractional calculus and their application to distributed systems, J. Franklin Instt., vol. **303**, no. 3, pp. 267–284, 1977.
6. Sannuti, P., Analysis and synthesis of dynamic systems via block-pulse functions, Proc. IEE, vol. **124**, no. 6, pp. 569–571, 1977.
7. Rao, G. P. and Srinivasan, T., Analysis and synthesis of dynamic systems containing time delays via block pulse functions, Proc. IEE, vol. **125**, no. 9, pp. 1064–1068, 1978.
8. Jiang, Z. H. and Schaufelberger, W., Recursive block pulse function method, in *Identification of continuous-time systems*,

N. K. Sinha and G. P. Rao (ed.), Kluwer Academic Publishers, Dordrecht, pp. 205–226, 1991.

9. Deb, A., Sarkar, G. and Sen, S. K., Linearly pulse-width modulated block pulse functions and their application to linear SISO feedback control system identification, IEE Proc. Control Theory and Appl., vol. **142**, no. 1, pp. 44–50, 1995.

10. Liou, C. T. and Chou, Y. S., Piecewise linear polynomial functions and their applications to analysis and parameter identification, Int. J. Syst. Sci., vol. **18**, no. 10, pp. 1919–1929, 1987.

11. Deb, A., Sarkar, G., Bhattacharjee, M. and Sen, S. K., A new set of piecewise constant orthogonal functions for the analysis of linear SISO systems with sample-and-hold, J. Franklin Instt., vol. **335B**, no. 2, pp. 333–358, 1998.

12. Deb, A., Sarkar, G. and Dasgupta, A., A complementary pair of orthogonal triangular function sets and its application to the analysis of SISO control systems, J. Instt. Engrs (India), vol. **84**, December, pp. 120–129, 2003.

Chapter 5

Convolution Process in Triangular Function Domain and Its Use in SISO Control System Analysis

Having established the orthogonal triangular function sets, it is worthwhile to determine the convolution of two real-valued functions, say, $f_1(t)$ and $f_2(t)$, in triangular function domain.

In control system analysis, the well-known relation [1] involving the input and the output of a linear time invariant system is given by

$$C(s) = G(s)R(s) \qquad (5.1)$$

where, $C(s)$ is the Laplace transform of the output; $G(s)$ is the transfer function of the plant; $R(s)$ is the Laplace transform of the input; and s is the Laplace operator.

In time domain, equation (5.1) takes the form

$$c(t) = \int_0^\infty g(\tau)r(t-\tau)d\tau \qquad (5.2)$$

That is, the output $c(t)$ is determined by evaluating the convolution integral on the RHS of equation (5.2), where it has been assumed that the integral exists.

Evaluation of this integral is frequently needed in the analysis of control systems. In what follows, such an integral is evaluated in its general form in triangular function domain and the results are used to determine the output of a single input single output (SISO) linear control system.

5.1 Convolution Integral

Convolution [2] of two functions is a significant physical concept in many diverse scientific fields. However, as in the case

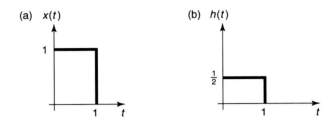

Figure 5.1. *Two typical waveforms for convolution.*

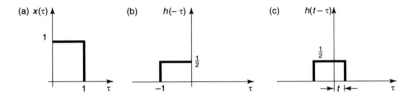

Figure 5.2. *Graphical illustration of folding and displacement operations.*

of many important mathematical relationships, the convolution integral does not readily unveil itself as to its true implications. The convolution integral is given by

$$y(t) = \int_{-\infty}^{\infty} x(\tau)h(t - \tau)\mathrm{d}\tau = x(t) * h(t) \qquad (5.3)$$

where, $*$ indicates convolution.

Let $x(t)$ and $h(t)$ be two time functions, as represented in Fig. 5.1(a) and (b), respectively.

To evaluate equation (5.3), functions $x(\tau)$ and $h(t - \tau)$ are required. The functions $x(\tau)$ and $h(\tau)$ are obtained by changing the variable t to the variable τ. $h(-\tau)$ is the image of $h(\tau)$ about the ordinate axis and $h(t - \tau)$ is the function $h(-\tau)$ shifted by the quantity t. Functions $x(\tau), h(-\tau)$, and $h(t - \tau)$ are shown in Fig. 5.2. The result of convolution of $x(t)$ and $h(t)$, as per equation (5.3), is the triangular function shown in Fig. 5.3.

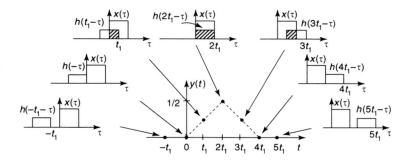

Figure 5.3. *Graphical example of convolution [2].*

We can summarize the steps for convolution as:

(i) *Folding.* Take the mirror image of $h(\tau)$ about the ordinate axis as shown in Fig. 5.2(b).

(ii) *Displacement.* Shift $h(-\tau)$ by the amount t as shown in Fig. 5.2(c).

(iii) *Multiplication.* Multiply the shifted function $h(t-\tau)$ by $x(T)$.

(iv) *Integration.* The area under the product of $h(t-\tau)$ and $x(\tau)$ is the value of the convolution at time t.

5.2 Convolution in Triangular Function Domain [3]

If two functions $f_1(t)$ and $f_2(t)$ are defined and continuous for all t, then their convolution $[f_1(t) * f_2(t)]$ is given by the well-known integral

$$\text{CV12}(t) \triangleq \int_{-\infty}^{\infty} f_1(\tau)f_2(t-\tau)d\tau \qquad (5.4)$$

For convolution of two functions, one function may be considered to be the *static function* or, say, STF (in our case, $f_1(t)$), and the other can be termed as the *scanning function* or, say, SCF (in our case $f_2(t)$).

If $f_1(t)$ and $f_2(t)$ are to be convolved in triangular function domain, the first task is to expand the functions $f_1(t)$ and $f_2(t)$ in terms of triangular functions. Then, we need the results of

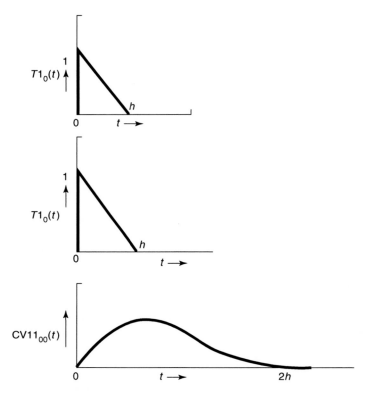

Figure 5.4. *Convolution of two members of the LHTF, $T1_0(t)$ and $T1_0(t)$.*

convolution of component triangular functions and use the same to determine the final result, namely $CV12(t)$.

5.2.1 Convolution of basis triangular function components

In the following, convolution results for the convolution of few typical members of the triangular function sets are shown. Figure 5.4 shows a typical curve of $CV11_{00}(t)$, which is given by

$$CV11_{00}(t) = T1_0(t) * T1_0(t)$$

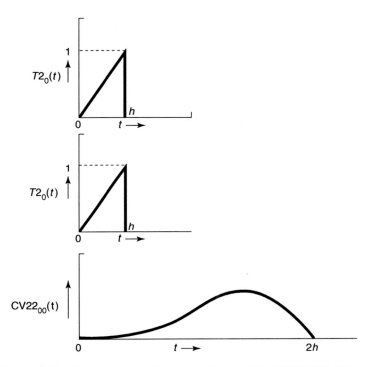

Figure 5.5. *Convolution of two members of the RHTF, $T2_0(t)$ and $T2_0(t)$.*

Similarly, two other convolution results are plotted in Figs 5.5 and 5.6. These functions are given by

$$CV22_{00}(t) = T2_0(t) * T2_0(t)$$

and

$$CV12_{00}(t) = T1_0(t) * T2_0(t)$$

When we express two time functions $f_1(t)$ and $f_2(t)$ in triangular function domain and try to compute the result of their convolution, four types of possibilities arise, with respect to triangular function components:

(i) Convolution of two members of the LHTF set **T1**.
(ii) Convolution of two members of the RHTF set **T2**.

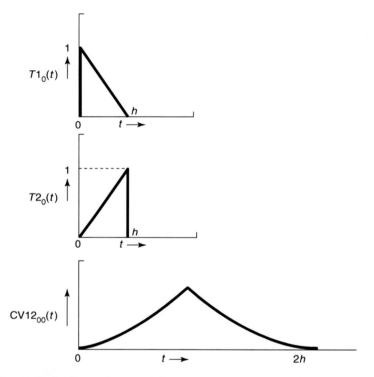

Figure 5.6. *Convolution of first member of the LHTF set and first member of the RHTF set, $T1_0(t)$ and $T2_0(t)$.*

   (iii) Convolution of a member of the LHTF set **T1** with another member of the RHTF set **T2**.

   (iv) Convolution of a member of the RHTF set **T2** with another member of the LHTF set **T1**.

It is to be noted that the result of (iii) and (iv) are essentially the same.

In the expansion of $f_1(t)$, let $T1_p(t)$ be the pth member of the LHTF set having a coefficient $c1_p$, and in the expansion of $f_2(t)$, let $T1_q(t)$ be the qth member of the LHTF set with coefficient $c2_p$.

Similarly, for the same expansion, let $T2_p(t)$ and $T2_q(t)$ be the pth and qth members of the RHTF set having coefficients $d1_p$ and $d2_p$, respectively.

Then, in general, we present the following algorithm for convolution for better clarity.

(i) Express $f_1(t)$ in terms of TF.

(ii) Express $f_2(t)$ in terms of TF.

(iii) Convolve $T1_p(t)$ and $T1_q(t)$ having coefficients $c1_p$ and $c2_q$, respectively.

(iv) Express the result in TF domain.

(v) Convolve $T2_p(t)$ and $T2_q(t)$ having coefficients $d1_p$ and $d2_q$, respectively.

(vi) Express the result in TF domain.

(vii) Convolve $T1_p(t)$ and $T2_q(t)$ in the same fashion as in step (iii).

(viii) Express the result in TF domain.

(ix) Convolve $T2_p(t)$ and $T1_q(t)$ in a manner similar to step (vii).

(x) Express the result in TF domain.

(xi) Use the results of steps (iv), (vi), (viii), and (x) to determine the final result CV12 in TF domain.

5.2.2 Convolution of $T1_p$ and $T1_q$ in TF domain

The convolving functions $T1_p$ and $T1_q$, having amplitudes $c1_p$, and $c2_q$, respectively, are shown in Fig. 5.7. Since, for generality, the functions have different amplitudes and delays, the convolution result $CV11_{pq}(t)$ is given by

$$CV11_{pq}(t) = \begin{cases} 0 & \text{for } 0 \le t < (p+q)h \\ c1_p c2_q[t - t^2/h + t^3/(6h^2)] \\ \quad \text{for } (p+q)h \le t < (p+q+1)h \\ c1_p c2_q[4h/3 - 2t + t^2/h - t^3/(6h^2)] \\ \quad \text{for } (p+q+1)h \le t < (p+q+2)h \end{cases}$$

$$(5.5)$$

The resulting function is shown in Fig. 5.8.

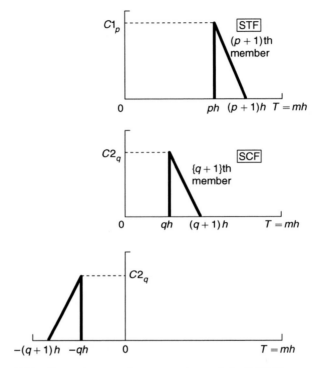

Figure 5.7. *Convolution of two members of the LHTF, namely $T1_p$ and $T1_q$.*

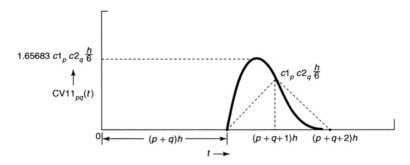

Figure 5.8. *A typical curve showing the result of convolution of two members of LHTF (having unit amplitude each, i.e., $c1_p = c2_q = 1$) and its TF domain approximation.*

To express the result in TF domain, we need to take samples of $CV11_{pq}(t)$ at $(p+q)h$, $(p+q+1)h$ and $(p+q+2)h$. Thus, from Fig. 5.8, we have

$$
CV11_{pq}(t) = \left[\underbrace{0\ 0\ \ldots\ 0}_{\substack{(p+q)\\ \text{zeros}}}\ 0\ c1_pc2_q\frac{h}{6}\ \underbrace{0\ \ldots\ 0\ 0}_{\substack{\{m-(p+q+2)\}\\ \text{zeros}}} \right] \mathbf{T1}_{(\mathbf{m})}
$$

$$
+ \left[\underbrace{0\ 0\ \ldots\ 0}_{\substack{(p+q+1)\\ \text{zeros}}}\ c1_pc2_q\frac{h}{6}\ 0\ \underbrace{0\ \ldots\ 0\ 0}_{\substack{\{m-(p+q+3)\}\\ \text{zeros}}} \right] \mathbf{T2}_{(\mathbf{m})}
$$

$$(5.6)$$

5.2.3 Convolution of other triangular function components in TF domain

Following the procedure outlined above and the *Algorithm*, the results of convolution of other possible combinations of triangular function components, with different delays and amplitudes, are

(i) steps (v) and (vi) of the Algorithm

$$
CV22_{pq}(t) = \left[\underbrace{0\ 0\ \ldots\ 0}_{\substack{(p+q)\\ \text{zeros}}}\ 0\ d1_pd2_q\frac{h}{6}\ \underbrace{0\ \ldots\ 0\ 0}_{\substack{\{m-(p+q+2)\}\\ \text{zeros}}} \right] \mathbf{T1}_{(\mathbf{m})}
$$

$$
+ \left[\underbrace{0\ 0\ \ldots\ 0}_{\substack{(p+q+1)\\ \text{zeros}}}\ d1_pd2_q\frac{h}{6}\ 0\ \underbrace{0\ \ldots\ 0\ 0}_{\substack{\{m-(p+q+3)\}\\ \text{zeros}}} \right] \mathbf{T2}_{(\mathbf{m})}
$$

$$(5.7)$$

(ii) steps (vii) and (viii) of the Algorithm

$$
CV12_{pq}(t) = \left[\underbrace{0\ 0\ \ldots\ 0}_{\substack{(p+q) \\ \text{zeros}}}\ 0\ c1_p d2_q \frac{h}{3}\ \underbrace{0\ \ldots\ 0\ 0}_{\substack{\{m-(p+q+2)\} \\ \text{zeros}}} \right] \mathbf{T1}_{(m)}
$$

$$
+ \left[\underbrace{0\ 0\ \ldots\ 0}_{\substack{(p+q+1) \\ \text{zeros}}}\ c1_p d2_q \frac{h}{3}\ 0\ \underbrace{0\ \ldots\ 0\ 0}_{\substack{\{m-(p+q+3)\} \\ \text{zeros}}} \right] \mathbf{T2}_{(m)}
$$

$$(5.8)$$

(iii) steps (ix) and (x) of the Algorithm

$$
CV21_{pq}(t) = \left[\underbrace{0\ 0\ \ldots\ 0}_{\substack{(p+q) \\ \text{zeros}}}\ 0\ d1_p c2_q \frac{h}{3}\ \underbrace{0\ \ldots\ 0\ 0}_{\substack{\{m-(p+q+2)\} \\ \text{zeros}}} \right] \mathbf{T1}_{(m)}
$$

$$
+ \left[\underbrace{0\ 0\ \ldots\ 0}_{\substack{(p+q+1) \\ \text{zeros}}}\ d1_p c2_q \frac{h}{3}\ 0\ \underbrace{0\ \ldots\ 0\ 0}_{\substack{\{m-(p+q+3)\} \\ \text{zeros}}} \right] \mathbf{T2}_{(m)}
$$

$$(5.9)$$

5.2.4 Convolution of two triangular function trains in TF domain

Now, we extend our ideas to convolution of two triangular function trains comprised of four component functions ($m = 4$), having different amplitudes.

To do this, we take two trains of **T1** (say, **T1**$_1$ and **T1**$_2$) basis as shown in Fig. 5.9. Following the approach described earlier, the result of convolution, expressed in matrix form, is

$$
CV11_{(4)}(t) = \frac{h}{6}[c2_0\ c2_1\ c2_2\ c2_3][0\ c1_0\ c1_1\ c1_2]\mathbf{T1}_{(4)}
$$

$$
+ \frac{h}{6}[c2_0\ c2_1\ c2_2\ c2_3][c1_0\ c1_1\ c1_2\ c1_3]\mathbf{T2}_{(4)}
$$

$$(5.10)$$

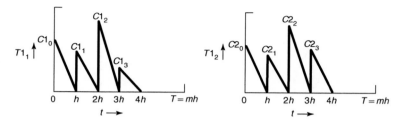

Figure 5.9. *Two trains of triangular functions, $T1_1$ and $T1_2$, of the $T1$ basis.*

where

$$[\![a \ b \ c]\!] = \begin{bmatrix} a & b & c \\ 0 & a & b \\ 0 & 0 & a \end{bmatrix}$$

With other combinations of triangular function trains (e.g., **T1** ∗ **T2**, **T2** ∗ **T1**, and **T2** ∗ **T2**) similar convolution results may be obtained.

5.3 Convolution of Two Time Functions in TF Domain

Referring to equation (2.34), we now express $f_1(t)$ and $f_2(t)$ in TF domain having m basis functions. Thus

$$\begin{aligned}
f_1(t) &\approx [c1_0 \ c1_1 \ c1_2 \ \ldots \ c1_{(m-1)}]\mathbf{T1_{(m)}} \\
&\quad + [d1_0 \ d1_1 \ d1_2 \ \ldots \ d1_{(m-1)}]\mathbf{T2_{(m)}} \\
&\triangleq \mathbf{C1^T T1_{(m)}} + \mathbf{D1^T T2_{(m)}} \qquad (5.11) \\
f_2(t) &\approx [c2_0 \ c2_1 \ c2_2 \ \ldots \ c2_{(m-1)}]\mathbf{T1_{(m)}} \\
&\quad + [d2_0 \ d2_1 \ d2_2 \ \ldots \ d2_{(m-1)}]\mathbf{T2_{(m)}} \\
&\triangleq \mathbf{C2^T T1_{(m)}} + \mathbf{D2^T T2_{(m)}} \qquad (5.12)
\end{aligned}$$

In Fig. 5.10, the typical vectors **C1**, **D1**, **C2**, and **D2** are shown. Convolution of $f_1(t)$ and $f_2(t)$ of equations (5.11) and (5.12) will involve convolution of four sets of triangular function trains. The component results thus obtained are to be conglomerated (step (xi) of Algorithm) to compute the final result That is

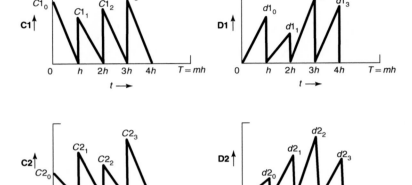

Figure 5.10. *Time diagram of the typical vectors* $C1, D1, C2,$ *and* $D2$ *resulting from TF domain expansion of the functions* $f_1(t)$ *and* $f_2(t)$.

$$CV12(t) \approx (\mathbf{C1}^T\mathbf{T1} * \mathbf{C2}^T\mathbf{T1}) + (\mathbf{C1}^T\mathbf{T1} * \mathbf{D2}^T\mathbf{T2})$$
$$+ (\mathbf{D1}^T\mathbf{T2} * \mathbf{C2}^T\mathbf{T1}) + (\mathbf{D1}^T\mathbf{T2} * \mathbf{D2}^T\mathbf{T2})$$
$$= \mathbf{C}^T\mathbf{T1} + \mathbf{D}^T\mathbf{T2}$$

As elaborated in the foregoing, the result of convolution $CV12(t)$, in TF domain, is given by

$$CV12(t) \approx \frac{h}{6}[c2_0 \; c2_1 \; c2_2 \; c2_3][\![0 \; c1_0 \; c1_1 \; c1_2]\!]\mathbf{T1}_{(4)}$$
$$+ \frac{h}{6}[c2_0 \; c2_1 \; c2_2 \; c2_3][\![c1_0 \; c1_1 \; c1_2 \; c1_3]\!]\mathbf{T2}_{(4)}$$
$$+ \frac{h}{3}[d2_0 \; d2_1 \; d2_2 \; d2_3][\![0 \; c1_0 \; c1_1 \; c1_2]\!]\mathbf{T1}_{(4)}$$
$$+ \frac{h}{3}[d2_0 \; d2_1 \; d2_2 \; d2_3][\![c1_0 \; c1_1 \; c1_2 \; c1_3]\!]\mathbf{T2}_{(4)}$$
$$+ \frac{h}{3}[c2_0 \; c2_1 \; c2_2 \; c2_3][\![0 \; d1_0 \; d1_1 \; d1_2]\!]\mathbf{T1}_{(4)}$$

$$+ \frac{h}{3}[c2_0 \; c2_1 \; c2_2 \; c2_3][\![d1_0 \; d1_1 \; d1_2 \; d1_3]\!]\mathbf{T2}_{(\mathbf{4})}$$

$$+ \frac{h}{6}[d2_0 \; d2_1 \; d2_2 \; d2_3][\![0 \; d1_0 \; d1_1 \; d1_2]\!]\mathbf{T1}_{(\mathbf{4})}$$

$$+ \frac{h}{6}[d2_0 \; d2_1 \; d2_2 \; d2_3][\![d1_0 \; d1_1 \; d1_2 \; d1_3]\!]\mathbf{T2}_{(\mathbf{4})}$$

$$(5.13)$$

Equation (5.13) is not general in the sense that it computes the convolution in TF domain for $m = 4$ only. However, it is an easy task to form the general equation by proper row and square matrices. That is, all matrices of order (1×4) of equation (5.13) is to be replaced by $(1 \times m)$ matrices, and all (4×4) matrices by $(m \times m)$ matrices.

Thus, the result of convolution in the TF domain for two functions having m-terms each, is

$CV12_{(t)}$

$$\approx \frac{h}{6}[c2_0 \; c2_1 \; \ldots \; c2_{(m-2)} \; c2_{(m-1)}][\![0 \; c1_0 \; c1_1 \; \ldots \; c1_{(m-2)}]\!]\mathbf{T1}_{(\mathbf{m})}$$

$$+ \frac{h}{6}[c2_0 \; c2_1 \; \ldots \; c2_{(m-2)} \; c2_{(m-1)}][\![c1_0 \; c1_1 \; \ldots \; c1_{(m-1)}]\!]\mathbf{T2}_{(\mathbf{m})}$$

$$+ \frac{h}{3}[d2_0 \; d2_1 \; \ldots \; d2_{(m-2)} \; d2_{(m-1)}][\![0 \; c1_0 \; c1_1 \; \ldots \; c1_{(m-2)}]\!]\mathbf{T1}_{(\mathbf{m})}$$

$$+ \frac{h}{3}[d2_0 \; d2_1 \; \ldots \; d2_{(m-2)} \; d2_{(m-1)}][\![\; c1_0 \; c1_1 \; \ldots \; c1_{(m-2)}]\!]\mathbf{T2}_{(\mathbf{m})}$$

$$+ \frac{h}{3}[c2_0 \; c2_1 \; \ldots \; c2_{(m-2)} \; c2_{(m-1)}][\![0 \; d1_0 \; d1_1 \; \ldots \; d1_{(m-2)}]\!]\mathbf{T1}_{(\mathbf{m})}$$

$$+ \frac{h}{3}[c2_0 \; c2_1 \; \ldots \; c2_{(m-2)} \; c2_{(m-1)}][\![d1_0 \; d1_1 \; \ldots \; d1_{(m-1)}]\!]\mathbf{T2}_{(\mathbf{m})}$$

$$+ \frac{h}{6}[d2_0 \; d2_1 \; \ldots \; d2_{(m-2)} \; d2_{(m-1)}][\![0 \; d1_0 \; d1_1 \; \ldots \; d1_{(m-2)}]\!]\mathbf{T1}_{(\mathbf{m})}$$

$$+ \frac{h}{6}[d2_0 \; d2_1 \; \ldots \; d2_{(m-2)} \; d2_{(m-1)}][\![d1_0 \; d1_1 \; \ldots \; d1_{(m-1)}]\!]\mathbf{T2}_{(\mathbf{m})}$$

$$(5.14)$$

5.4 Numerical Experiment

Consider a first-order plant having a transfer function [1]

$$G(s) = (s + 1)^{-1}$$

For n input $r(t) = u(t)$, the output is given by

$$C(s) = [s(s + 1)]^{-1}$$

Thus,

$$c(t) = r(t) * g(t)$$

where, $g(t) = \exp(-t)$, for $t \geq 0$.
Exact solution for plant response is

$$c(t) = 1 - \exp(-t), \quad \text{for } t \geq 0 \tag{5.15}$$

Triangular function expansion of $r(t)$ and $g(t)$, for $T = 1$ s and $m = 10$, following equation (2.35) yields

$$r(t) = [1 \ 1 \ 1 \ 1 \ 1 \ 1 \ 1 \ 1 \ 1 \ 1]\mathbf{T1}$$
$$+ [1 \ 1 \ 1 \ 1 \ 1 \ 1 \ 1 \ 1 \ 1 \ 1]\mathbf{T2}$$

$$g(t) \approx [1 \ 0.9048375 \ 0.8187308 \ 0.7408182 \ 0.6703200$$
$$0.6065307 \ 0.5488116 \ 0.4965853 \ 0.4493290$$
$$0.4065696]\mathbf{T1}$$
$$+ [0.9048375 \ 0.8187308 \ 0.7408182 \ 0.6703200$$
$$0.6065307 \ 0.5488116 \ 0.4965853 \ 0.4493290$$
$$0.4065696 \ 0.3678794]\mathbf{T2}$$

The output, according to equation (5.14), is

$$c(t) \approx [0 \ 0.0952419 \ 0.1814203 \ 0.2593978 \ 0.3299547$$
$$0.3937972 \ 0.4515643 \ 0.5038342 \ 0.5511299$$
$$0.5939248]\mathbf{T1}$$
$$+ [0.0952419 \ 0.1814203 \ 0.2593978 \ 0.3299547$$
$$0.3937972 \ 0.4515643 \ 0.5038342 \ 0.5511299$$
$$0.5939248 \ 0.6326473]\mathbf{T2} \tag{5.16}$$

However, direct TF domain expansion of $c(t)$ in (5.15), for $m = 10$ and $T = 1$ s, is

$$c(t) \approx [0 \quad 0.0951625 \quad 0.1812692 \quad 0.2591818 \quad 0.3296800$$
$$0.3934693 \quad 0.4511884 \quad 0.5034148 \quad 0.5506710$$
$$0.5934304]\mathbf{T1}$$
$$+ [0.0951625 \quad 0.1812692 \quad 0.2591818 \quad 0.3296800$$
$$0.3934693 \quad 0.4511884 \quad 0.5034148 \quad 0.5506710$$
$$0.5934304 \quad 0.6321206]\mathbf{T2} \tag{5.17}$$

Comparing equations (5.16) and (5.17), it is observed that the coefficients are very close indicating efficiency of the method.

5.5 Integral Squared Error (ISE) in TF Domain and Its Comparison with BPF Domain Solution

Let us consider $(m+1)$ sample points of the function $f(t)$, having a sampling period h, denoted by $f(ih), i = 0, 1, 2, \ldots, m$. Then piecewise linear representation of the function $f(t)$ by triangular functions is obtained simply by joining these sample points. The equation of one such straight line in the ith interval is given by

$$\hat{f}(t) = m_i t + f(ih) - im_i h \tag{5.18}$$

where,

$$m_i = \frac{f[(i+1)h] - f(ih)}{h} \tag{5.19}$$

Then mean integral squared error (MISE) in the ith interval is given by

$$[\epsilon_i]^2 \triangleq \frac{1}{h} \int_{ih}^{(i+1)h} [f(t) - \hat{f}(t)]^2 dt \tag{5.20}$$

To compute the MISE for the result obtained in equation (5.16), we naturally make use of equation (5.20). The coefficients of equation (5.16) form the function $\hat{f}(t)$, while $f(t)$ is the exact analytical expression of $c(t)$. Since the interval is 1 s, MISE and ISE are the same in this case.

Following the methodology delineated in Section 3.3, the MISE in the interval [0, 1) for $m = 10$, is found to be 2.5732038×10^{-7}.

A similar analysis is of special interest in the BPF domain to judge the efficiency of the TF domain solution against its BPF counterpart. Following Kwong and Chen [4], the BPF domain solution for the output of the same plant via convolution is

$$c(t) \approx [0.0475813 \quad 0.1382159 \quad 0.2202255 \quad 0.2944309$$
$$0.3615746 0.4223288 \quad 0.4773015 \quad 0.5270429$$
$$0.5720507 \quad 0.6127755] \Psi_{(10)}$$

Following equation (5.20), the MISE in the BPF domain is 3.6021721×10^{-4}.

Thus,

$$\frac{\text{MISE}_{\text{BPF}}}{\text{MISE}_{\text{TF}}} = 1399.8782551 \qquad (5.21)$$

This establishes the superiority of the TF domain analysis over the BPF domain.

5.6 Conclusion

The proposed triangular function sets have been applied to determine the convolution of two triangular function components as well as triangular function trains. The result has been used to compute the output of a linear SISO control system via an operational technique. The estimated MISE (vide Section 3.3) is found to be much less than the alternative staircase solution obtained from BPF domain analysis.

The analysis technique presented herein is based upon nonoptimal determination of the expansion coefficients to enjoy the advantages of using function samples as the coefficients like in Refs [5,6]. A detailed comparison of optimal and nonoptimal expansion coefficients with related MISE of a few standard functions (using triangular functions, block pulse functions, and sample-and-hold functions) has been presented in Section 3.4.

It was noted that, though MISE is larger for nonoptimal TF domain analysis, advantages of using function samples outweigh the slight loss in accuracy.

References

1. Ogata, K., *Modern control engineering* (3rd ed.), Prentice-Hall of India Ltd., New Delhi, 1997.
2. Brigham, E. O., *The fast Fourier transform and its applications*, Prentice-Hall International Inc., New Jersey, 1988.
3. Deb, A., Sarkar G. and Dasgupta, A., A complementary pair of orthogonal triangular function sets and its application to the analysis of SISO control systems, J. Instt. Engrs (India), vol. **84**, December, pp. 120–129, 2003.
4. Kwong, C. P. and Chen, C. F., Linear feedback system identification via block pulse functions, Int. J. Syst. Sci., vol. **12**, no. 5, pp. 635–642, 1981.
5. Deb, A., Sarkar, G., Bhattacharjee, M. and Sen, S. K., Analysis of linear discrete SISO control systems via a set of delta functions, IEE Proc., Control Theory and Appl., vol. **143**, no. 6, pp. 514–518, 1996.
6. Deb, A., Sarkar, G., Bhattacharjee, M. and Sen, S. K., A new set of piecewise constant orthogonal functions for the analysis of linear SISO systems with sample-and-hold, J. Franklin Instt., vol. **335B**, no. 2, pp. 333–358, 1998.

Chapter 6

Identification of SISO Control Systems via State Space Approach

System identification [1–4] is a common problem encountered in the design of control systems. The unknown components, usually the plant under control, are assumed to be described satisfactorily by its respective models. Then, the problem of identification is the characterization of the assumed model based on some observations or measurements. It is well known that one may set up more than one model for a dynamic system, and in control system design the choice of the most suitable one depends heavily on the design method [3] being used. For example, in linear control problems, a parametric model such as a state model is usually adopted in modern design. While in classical design, a nonparametric model such as an impulse response function is more appropriate [1]. Kwong and Chen [5], in their work, presented a method based upon block pulse functions (BPF) to identify an unknown plant modeled by impulse–response.

The basic equation which relates input–output of a control system in Laplace domain is well known and given by

$$C(S) = G(S)R(S) \tag{6.1}$$

or

$$G(S) = \frac{C(S)}{R(S)} \tag{6.2}$$

where $C(S)$ is the Laplace transformed output. $G(S)$ is the Transfer function of the system and $R(S)$ is the Laplace transformed input to the system.

The problem of identification is basically to determine $G(S)$ from equation (6.2).

$G(S)$ may be known via Markov parameters [2] of the system or else $G(S)$ may be computed by way of determining $g(t)$, which is the impulse response [6] of the system. Kwong and Chen [5] employed BPF to identify the plant of a feedback control system in the form of BPF expansion of the impulse response $g(t)$.

Another common way of identifying a system is to determine its state matrix **A**. The state equation of a linear control system is given by

$$\dot{\mathbf{x}}(\mathbf{t}) = \mathbf{A}\mathbf{x}(\mathbf{t}) + \mathbf{B}\mathbf{u}(\mathbf{t}) \tag{6.3}$$

where, $\mathbf{x(t)}$ is an n-state vector; **A**, state matrix of order n; **B**, input matrix of order $n \times 1$; and $\mathbf{u(t)}$, forcing function.

Identification of a system modeled by equation (6.3) will essentially mean determination of the state matrix **A**. Walsh functions were employed for system identification [7–10] in the initial stage of development of the piecewise constant orthogonal function family. Later, BPF [11–15], variants of BPF [16,17] and many other related functions [18–20] were employed for system identification. Linearly pulse-width modulated BPF [17] proved to be more efficient than the method suggested by Kwong and Chen [5] based upon BPFs.

In the following, we have employed the triangular function (TF) approach to identify a linear control system described by equation (6.3).

6.1 System Identification via State Space Approach

Suppose we are interested in realizing the system described by equation (6.3).

Let, each state $x_1, x_2, \ldots, x_n$ of $\mathbf{x(t)}$ be expanded into an m-term TF series. Then according to equation (2.34), we can write

$$\mathbf{x(t)} = \mathbf{ET1} + \mathbf{FT2} \tag{6.4}$$

where

$$\mathbf{E}^\mathrm{T} = [\mathbf{e_1} \quad \mathbf{e_2} \quad \cdots \quad \cdots \quad \mathbf{e_i} \quad \cdots \quad \cdots \quad \mathbf{e_{(n-1)}} \quad \mathbf{e_n}]_{m \times n} \tag{6.5}$$

$$\mathbf{F}^T = [\mathbf{f_1} \quad \mathbf{f_2} \quad \ldots \quad \ldots \quad \mathbf{f_i} \quad \ldots \quad \ldots \quad \mathbf{f_{(n-1)}} \quad \mathbf{f_n}]_{m \times n}$$
(6.6)

where, $\mathbf{e_i^T} = [e_{i0} \quad e_{i1} \quad e_{i2} \quad \ldots \quad e_{ij} \ldots e_{i(m-2)} \quad e_{i(m-1)}]$ and $\mathbf{f_i^T} = [f_{i0} \quad f_{i1} \quad f_{i2} \quad \ldots \quad f_{ij} \ldots f_{i(m-2)} \quad f_{i(m-1)}]$.

Similarly, the forcing function $u(t)$ is expanded into an m-term TF series given by

$$u(t) = \mathbf{H}^T\mathbf{T1} + \mathbf{K}^T\mathbf{T2}$$
(6.7)

where, $\mathbf{H}^T = [h_0 \quad h_1 \quad h_2 \quad \ldots \quad h_i \quad \ldots \quad h_{(m-2)} \quad h_{(m-1)}]$ and $\mathbf{K}^T = [k_0 \quad k_1 \quad k_2 \quad \ldots \quad k_i \quad \ldots \quad k_{(m-2)} \quad k_{(m-1)}]$, $\mathbf{H}$ and $\mathbf{K}$ are $m \times 1$ matrices, $h_0, h_1, \ldots, h_{(m-1)}, h_{(m)}$ are samples of the forcing function $u(t)$, and $h_{(i+1)} = k_i$, $i = 0, 1, 2, \ldots, (m-1)$.

Integrating both sides of equation (6.3), and using equations (3.5), (3.7), and the operational matrices **P1** and **P2**, we have

$$\int_0^t \dot{\mathbf{x}}(\tau)d\tau = \mathbf{A} \int_0^t \mathbf{x}(\tau)d\tau + \mathbf{B} \int_0^t \mathbf{u}(\tau)d\tau$$

$$= \mathbf{A} \int_0^t (\mathbf{ET1} + \mathbf{FT2})d\tau + \mathbf{B} \int_0^t (\mathbf{H}^T\mathbf{T1} + \mathbf{K}^T\mathbf{T2})d\tau$$

Integrating and using the results of equations (3.5) and (3.7), we get

$$\mathbf{x(t)} - \mathbf{x(0)} = \mathbf{AE(P1T1 + P2T2)} + \mathbf{AF(P1T1 + P2T2)}$$
$$+ \mathbf{BH}^T\mathbf{(P1T1 + P2T2)} + \mathbf{BK}^T\mathbf{(P1T1 + P2T2)}$$

or

$$\mathbf{E'T1} + \mathbf{F'T2} = [\mathbf{A(E + F)P1} + \mathbf{B(H^T + K^T)P1}]\mathbf{T1}$$
$$+ [\mathbf{A(E + F)P2} + \mathbf{B(H^T + K^T)P2}]\mathbf{T2} \quad (6.8)$$

where, $\mathbf{E'}$ and $\mathbf{F'}$ are formed by subtracting $\mathbf{x(0)}$ from matrices $\mathbf{E}$ and $\mathbf{F}$, respectively.

Segregating the coefficients of **T1** and **T2** in equation (6.8), we can write

$$\mathbf{E'T1} = [\mathbf{A(E + F)P1} + \mathbf{B(H^T + K^T)P1}]\mathbf{T1}$$
(6.9)

$$\mathbf{F'T2} = [\mathbf{A(E + F)P2} + \mathbf{B(H^T + K^T)P2}]\mathbf{T2}$$
(6.10)

Now the vectors **A** and **B** may be solved using either of equations (6.9) and (6.10).

Considering equation (6.10), it is noted that the equation holds for any instant of time, namely t. Since total number of unknowns is $(n+1)$, let us take $(n+1)$ samples of the component functions of the basis vector **T2** at time instants $t_1, t_2, \ldots, t(n+1)$ to form $(n+1)$ equations.

Thus, using $(n+1)$ samples and rearranging (6.10), we can write

$$[\mathbf{F}']_{n \times m}[\mathbf{S}]_{m \times (n+1)}$$

$$= [\mathbf{A} \quad \mathbf{B}]_{n \times (n+1)} \begin{bmatrix} (\mathbf{E}+\mathbf{F})\mathbf{P2} \\ (\mathbf{H}^T + \mathbf{K}^T)\mathbf{P2} \end{bmatrix}_{(n+1) \times m} [\mathbf{S}]_{m \times (n+1)} \quad (6.11)$$

where,

$$\mathbf{S} = \begin{bmatrix} T2_0(t_1) & T2_0(t_2) & T2_0(t_3) & \ldots \\ T2_1(t_1) & T2_1(t_2) & T2_1(t_3) & \ldots \\ \ldots & \ldots & \ldots & \ldots \\ T2_{(m-2)}(t_1) & T2_{(m-2)}(t_2) & \ldots & \ldots \\ T2_{(m-1)}(t_1) & T2_{(m-1)}(t_2) & \ldots & \ldots \end{bmatrix}$$

$$\begin{bmatrix} T2_0(t_n) & T2_0(t_{(n+1)}) \\ T2_1(t_n) & T2_1(t_{(n+1)}) \\ \ldots & \ldots \\ T2_{(m-2)}(t_n) & T2_{(m-2)}(t_{(n+1)}) \\ T2_{(m-1)}(t_n) & T2_{(m-1)}(t_{n+1}) \end{bmatrix}_{m \times (n+1)} \quad (6.12)$$

or

$$[\mathbf{F}']_{n \times m}[\mathbf{S}]_{m \times (n+1)} = [\mathbf{A} \quad \mathbf{B}]_{n \times (n+1)}[\mathbf{Q}]_{(n+1) \times (n+1)},$$

where

$$[\mathbf{Q}]_{(n+1) \times (n+1)} = \begin{bmatrix} (\mathbf{E}+\mathbf{F})\mathbf{P2} \\ (\mathbf{H}^T + \mathbf{K}^T)\mathbf{P2} \end{bmatrix}_{(n+1) \times m} [\mathbf{S}]_{m \times (n+1)}$$

or

$$[\mathbf{A} \quad \mathbf{B}]_{n \times (n+1)} = [\mathbf{F}']_{n \times m}[\mathbf{S}]_{m \times (n+1)}[\mathbf{Q}]^{-1}_{(n+1) \times (n+1)} \quad (6.13)$$

Equation (6.13) gives the solution for the matrices **A** and **B**.

6.2 Numerical Example [6]

Consider the system

$$\dot{\mathbf{x}} = \mathbf{A}\mathbf{x} + \mathbf{B}u$$

or

$$\begin{bmatrix} \dot{x}1 \\ \dot{x}2 \end{bmatrix} = \begin{bmatrix} 0 & 1 \\ -2 & -3 \end{bmatrix} \begin{bmatrix} x1 \\ x2 \end{bmatrix} + \begin{bmatrix} 0 \\ 1 \end{bmatrix} u \qquad (6.14)$$

where u = unit step input

$$\left. \begin{array}{l} x1 = 0.5[1 + \exp(-2t)] - \exp(-t) \\ x2 = \exp(-t) - \exp(-2t) \\ \text{and} \quad x1(0) = x2(0) = 0 \end{array} \right\} \qquad (6.15)$$

Our task is to determine **A** and **B** knowing $x1$ and $x2$.

Let $T = 1$ s, $m = 4$, and $h = 0.25$. For TF domain analysis, we expand $x1$ and $x2$ as

$$x1 = [0 \quad 0.024464547 \quad 0.077409029 \quad 0.139198541]\mathbf{T1}$$
$$+ [0.024464547 \quad 0.077409029 \quad 0.139198541$$
$$0.199788212]\mathbf{T2} \qquad (6.16)$$
$$x2 = [0 \quad 0.172270119 \quad 0.238651216 \quad 0.249236404]\mathbf{T1}$$
$$+ [0.172270119 \quad 0.238651216 \quad 0.249236404$$
$$0.232544168]\mathbf{T2} \qquad (6.17)$$

From equations (6.16) and (6.17), we have

$$\mathbf{E} = \begin{bmatrix} 0 & 0.024464547 & 0.077409029 & 0.139198541 \\ 0 & 0.172270119 & 0.238651216 & 0.249236404 \end{bmatrix}$$

$$\mathbf{F} = \begin{bmatrix} 0.024464547 & 0.077409029 & 0.139198541 & 0.1997882120 \\ 0.172270119 & 0.238651216 & 0.249236404 & 0.232544168 \end{bmatrix}$$

Now the forcing function $u(t)$ is expressed in TF domain as

$$u(t) = [1 \quad 1 \quad 1 \quad 1]\mathbf{T1} + [1 \quad 1 \quad 1 \quad 1]\mathbf{T2}$$

Thus, referring to equation (6.7)

$$\mathbf{H}^T = [1 \quad 1 \quad 1 \quad 1] \quad \text{and} \quad \mathbf{K}^T = [1 \quad 1 \quad 1 \quad 1]$$

Now, knowing $\mathbf{x(0)}$, $\mathbf{E}$, $\mathbf{F}$, $\mathbf{H}$, $\mathbf{K}$, the operational matrix $\mathbf{P2}$, and the rectangular matrix $\mathbf{S}$, we can solve for the unknown matrices $\mathbf{A}$ and $\mathbf{B}$ from equation (6.13). Solving, we get

$$\mathbf{A} = \begin{bmatrix} -0.030290911 & 0.9645216 \\ -1.9290484 & -2.923875 \end{bmatrix} \quad \text{and}$$

$$\mathbf{B} = \begin{bmatrix} 0.01514958 \\ 0.9645253 \end{bmatrix} \tag{6.18}$$

It is noted that the results are close to the actual solution. If we use 10 triangular function components, i.e., $m = 10$, then

$$\mathbf{A} = \begin{bmatrix} -0.00494126 & 0.9941937 \\ -1.988331 & -2.9875679 \end{bmatrix} \quad \text{and}$$

$$\mathbf{B} = \begin{bmatrix} 0.00248706 \\ 0.99419350 \end{bmatrix} \tag{6.19}$$

It is seen that the accuracy of results has increased significantly.

Now to compare the above results with Walsh or BPF analysis, we use the following equation [7] to compute the state matrix $\mathbf{A}$, given by

$$\mathbf{A} = \{\mathbf{C} - [\mathbf{x(0)}, \quad 0, \quad 0, \quad 0, \quad 0 \quad \cdots \quad 0]\} \boldsymbol{\Psi} \{\mathbf{CP_\Psi}\}^{-1}$$

where, $\mathbf{C}$ is the matrix formed by BPF expansion of the state vector $\mathbf{x(t)}$. $\mathbf{P}$ is the operational matrix for integration in the BPF domain, $\boldsymbol{\Psi}$ and $\mathbf{x(0)}$ is the initial condition matrix for the state vector $\mathbf{x(t)}$. All matrices and vectors of the above equation are of proper dimensions.

BPF expansion of the states $x1(t)$ and $x2(t)$ are given by

$$x1 = [0.008672474 \quad 0.049570713 \quad 0.108092844$$
$$0.169846430] \boldsymbol{\Psi}_{(4)}$$
$$x2 = [0.097858183 \quad 0.211778059 \quad 0.247157886$$
$$0.242358714] \boldsymbol{\Psi}_{(4)}$$

For $T = 1$ s, $m = 4$, and $h = 0.25$, the BPF domain solution for **A** and **B** is

$$\mathbf{A} = \begin{bmatrix} -2.3123791 & 2.4841893 \\ 5.0038400 & -7.5405643 \end{bmatrix} \quad \mathbf{B} = \begin{bmatrix} -0.1536644 \\ 1.4773757 \end{bmatrix}$$

$$(6.20)$$

Again, for $T = 1$ s, we identify the system using 10 block pulse terms (i.e., $m = 10$). The solution for **A** and **B** is

$$\mathbf{A} = \begin{bmatrix} -6.5319639 & 2.1878439 \\ 17.6316229 & -6.5756492 \end{bmatrix} \quad \mathbf{B} = \begin{bmatrix} -0.058012091 \\ 1.1760535 \end{bmatrix}$$

$$(6.21)$$

It is observed that the results have deteriorated beyond recognition indicating simple failure of the BPF/Walsh domain approach for solving identification problems via state space.

Now, we analyze the same system via nonoptimal BPF domain following the rule described in Section 2.2.

In this domain, for $T = 1$ s, $m = 4$, and $h = 0.25$, the states are available as equidistant samples and the state vectors expressed via nonoptimal BPF domain are given by

$$x1 = [0.012232273 \quad 0.050936788 \quad 0.108303785$$
$$0.169493377]\Psi'_{(4)}$$
$$x2 = [0.086135059 \quad 0.205460667 \quad 0.243943810$$
$$0.240890286]\Psi'_{(4)} \tag{6.22}$$

where $\Psi'_{(4)}$ is the nonoptimal BPF vector of order 4.

The matrices **A** and **B** are obtained as

$$\mathbf{A} = \begin{bmatrix} -0.030290918 & 0.9645216 \\ -1.9290485 & -2.9238753 \end{bmatrix} \quad \text{and}$$

$$\mathbf{B} = \begin{bmatrix} 0.015149589 \\ 0.96452530 \end{bmatrix} \tag{6.23}$$

For $T = 1$ s, $m = 10$, and $h = 0.1$, $\mathbf{A}$ and $\mathbf{B}$ are found to be

$$\mathbf{A} = \begin{bmatrix} -0.004941243 & 0.9941938 \\ -1.9883312 & -2.9875679 \end{bmatrix} \text{ and}$$

$$\mathbf{B} = \begin{bmatrix} 0.002487069 \\ 0.99419350 \end{bmatrix} \tag{6.24}$$

It is noted that the results obtained via nonoptimal BPF domain match exceptionally with those obtained through TF domain analysis. This is not at all surprising, because any nonoptimal BPF coefficient is nothing but the *average of two consecutive TF coefficients*. In spite of this surprising corroboration, TF domain approach does not lose its merit due to the fact that this approach incurs much lower integral squared error (ISE) while approximating square integrable functions of Lebesgue measure.

6.3 Conclusion

In the present work, with the help of the operational matrices **P1** and **P2**, the newly proposed triangular function sets have been applied to the identification of linear SISO dynamic systems via state space approach and results obtained matched nicely with the exact solution. A detailed study of the representational error in TF domain has already been presented in Section 3.3.

It was observed that, the BPF (or, Walsh) domain technique failed to identify the system completely even with $m = 10$ in an interval of [0, 1)s. However, nonoptimal BPF approach, which is mathematically equivalent to discrete Walsh function technique mentioned in Section 2.2, yields results as good as those obtained via TF domain method. This surprising superiority of nonoptimal BPF over optimal BPF for solving identification problems is a fact that warrants deeper exploration.

As usual, the analysis technique presented herein is based upon nonoptimal determination of the expansion coefficients obtained from function samples.

References

1. Eykhoff, P., *System identification: parameter and state estimation*, John Wiley & Sons., London, 1974.
2. Chen, C. T., *Linear system theory and design*, Holt Rinehart and Winston, Holt-Saunders, Japan, 1984.
3. Unbehauen, H. and Rao, G. P., *Identification of continuous systems*, North-Holland, Amsterdam, 1987.
4. Ljung, L., *System identification: theory for the user*, Prentice-Hall Inc., New Jersey, 1985.
5. Kwong, C. P. and Chen, C. F., Linear feedback system identification via block-pulse functions, Int. J. Syst. Sci., vol. **12**, no. 5, pp. 635–642, 1981.
6. Ogata, K., *Modern control engineering* (3rd ed.), Prentice-Hall of India Ltd., New Delhi, 1997.
7. Chen, C. F. and Hsiao, C. H., Time domain synthesis via Walsh functions, Proc. IEE, vol. **122**, no. 5, pp. 565–570, 1975.
8. Rao, G. P. and Sivakumar, L., System identification via Walsh functions, Proc. IEE, vol. **122**, no. 10, pp. 1160–1161. 1975.
9. Karanam, V. R., Frick, P. A. and Mohler, R. R., Bilinear system identification by Walsh functions, IEEE Trans. Automatic Control, vol. **AC–23**, no. 4, pp. 709–713, 1978.
10. Rao, G. P. and Sivakumar, L., Transfer function matrix identification in MIMO systems via Walsh functions, Proc. IEEE, vol. **69**, no. 4, pp. 465–466, 1981.
11. Sannuti, P., Analysis and synthesis of dynamic systems via block-pulse functions, Proc. IEE, vol. **124**, no. 6, pp. 569–571, 1977.
12. Wang, Chi-Hsu, Generalized block pulse operational matrices and their applications to operational calculus, Int. J. Control, vol. **36**, no. 1, pp. 67–76, 1982.
13. Jiang, Z. H. and Schaufelberger, W., Recursive block pulse function method, in *Identification of continuous-time systems*, N. K. Sinha and G. P. Rao (ed.), Kluwer Academic Publishers, Dordrecht, pp. 205–226, 1991.
14. Cheng, Bing and Hsu, Ning-Show, Single-input-single-output system identification via block-pulse functions, Int. J. Syst. Sci., vol. **13**, no. 6, pp. 697–702, 1982.
15. Rao, G. P. and Srinivasan, T., Analysis and synthesis of dynamic systems containing time delays via block pulse functions, Proc. IEE, vol. **125**, no. 9, pp. 1064–1068, 1978.

16. Wang, Chi-Hsu and Marleau, Richard S., System identification via generalized block pulse operational matrices, Int. J. Syst. Sci., vol. **16**, no. 11, pp. 1425–1430, 1985.
17. Deb, A., Sarkar, G. and Sen, S. K., Linearly pulse-width modulated block pulse functions and their application to linear SISO feedback control system identification, IEE Proc. Control Theory and Appl., vol. **142**, no. 1, pp. 44–50, 1995.
18. Liou, C. T. and Chou, Y. S., Piecewise linear polynomial functions and their applications to analysis and parameter identification, Int. J. Syst. Sci., vol. **18**, no. 10, pp. 1919–1929, 1987.
19. Wang, Maw-Ling, Yang, Shwu-Yien and Chang, Rong-Yeu, New approach for parameter identification via generalized orthogonal polynomials, Int. J. Syst. Sci., vol. **18**, no. 3, pp. 569–579, 1987.
20. Mohan, B. M. and Datta, K. B., Linear time-invariant distributed parameter system identification via orthogonal functions, Automatica, vol. **27**, no. 2, pp. 409–412, 1991.

Chapter 7

Solution of Integral Equations via Triangular Functions

In this chapter, efforts have been made to solve integral equations [1] using orthogonal triangular function (TF) sets. It is well known that, in control system analysis and synthesis, integral equations play an important role and often the control problem surmises to solving one or several integral equations. The proposed TFs, when used in a fashion similar to that of block pulse functions (BPF) [2,3], yield a piecewise linear solution of dynamic systems with less integral squared error (ISE).

The main objectives of the present work are:

(i) To solve Fredholm integral equation of the second kind [1] via TFs and compare the results with its BPF domain solution.

(ii) To solve Volterra integral equation of the second kind [1] via TFs and compare the results with its BPF domain solution.

The theory of the proposed TF method has been developed and then supported by several examples. Results are also compared with BPF domain solution with respect to ISE.

Solution of integral equations was of interest to the scientific community that is apparent from the year of publication of Ref. 1. With the advent of Walsh [4], block pulse [2,3], and related functions, interest of the researchers took a new turn and they thrived for finding out techniques that were computationally more attractive using new family of orthogonal functions. Such efforts with BPF continued to appear in journals as soon as Chen et al. formally introduced BPF in 1977 [5]. Kung and Chen [6],

Wang and Shih [7] used BPF to solve integral equations. Chang et al. [8] employed orthogonal polynominals for the past two decades, and yet another attractive method is presented in this chapter.

7.1 Solution of Integral Equations via Triangular Functions

7.1.1 Fredholm integral equation of the second kind

Such an equation is given by [1]

$$\alpha(x)y(x) = f(x) + \lambda \int_a^b K(x,t)y(t)dt \tag{7.1}$$

where, $\alpha(x)$, $f(x)$, and the kernel $K(x,t)$ are given. And a, b, and λ are constants.

Expanding $\alpha(x)$, $y(x)$, and $f(x)$ into corresponding m-set TF series over the relevant interval, we may write

$$\alpha(x) = \mathbf{A}^T \mathbf{T1}(\mathbf{x}) + \mathbf{B}^T \mathbf{T2}(\mathbf{x}) \tag{7.2}$$

$$y(x) = \mathbf{E}^T \mathbf{T1}(\mathbf{x}) + \mathbf{F}^T \mathbf{T2}(\mathbf{x}) \tag{7.3}$$

$$f(x) = \mathbf{G}^T \mathbf{T1}(\mathbf{x}) + \mathbf{H}^T \mathbf{T2}(\mathbf{x}) \tag{7.4}$$

where, $\mathbf{A}^T = [a_0 \quad a_1 \quad a_2 \quad \ldots \quad a_i \quad \ldots \quad a_{(m-1)}]$;
$\quad\quad \mathbf{B}^T = [b_0 \quad b_1 \quad b_2 \quad \ldots \quad b_i \quad \ldots \quad b_{(m-1)}]$;
$\quad\quad \mathbf{E}^T = [e_0 \quad e_1 \quad e_2 \quad \ldots \quad e_i \quad \ldots \quad e_{(m-1)}]$;
$\quad\quad \mathbf{F}^T = [f_0 \quad f_1 \quad f_2 \quad \ldots \quad f_i \quad \ldots \quad f_{(m-1)}]$;
$\quad\quad \mathbf{G}^T = [g_0 \quad g_1 \quad g_2 \quad \ldots \quad g_i \quad \ldots \quad g_{(m-1)}]$;
and $\mathbf{H}^T = [h_0 \quad h_1 \quad h_2 \quad \ldots \quad h_i \quad \ldots \quad h_{(m_1)}]$.

Obviously, our task is to determine the vectors $\mathbf{E}$ and $\mathbf{F}$.

From equations (7.2) and (7.3), using *Theorem 1* from Section 2.7, we have

$$\alpha(x)y(x) = \mathbf{A}^T * \mathbf{E}^T \mathbf{T1}(\mathbf{x}) + \mathbf{B}^T * \mathbf{F}^T \mathbf{T2}(\mathbf{x}) \tag{7.5}$$

Like all other terms of equation (7.1), $K(x, t)$ is also absolutely integrable in $[a, b]$ and it can be expressed in terms of TFs to yield

$$K(x, t) = \mathbf{R(t)}^T\mathbf{T1}(\mathbf{x}) + \mathbf{S(t)}^T\mathbf{T2}(\mathbf{x}) \qquad (7.6)$$

where,

$$\mathbf{R(t)} = \mathbf{M}^T\mathbf{T1(t)} + \mathbf{N}^T\mathbf{T2(t)} \qquad (7.7)$$
$$\mathbf{S(t)} = \mathbf{P}^T\mathbf{T1(t)} + \mathbf{Q}^T\mathbf{T2(t)} \qquad (7.8)$$

and **M**, **N**, **P**, and **Q** are square matrices of order m, each formed by the expansion coefficients of each element of the functions **R(t)** and **S(t)**.

Substituting equations (7.7) and (7.8) in equation (7.6), we have

$$\begin{aligned}
K(x, t) &= [\mathbf{M}^T\mathbf{T1(t)} + \mathbf{N}^T\mathbf{T2(t)}]^T\mathbf{T1}(\mathbf{x}) \\
&\quad + [\mathbf{P}^T\mathbf{T1(t)} + \mathbf{Q}^T\mathbf{T2(t)}]^T\mathbf{T2}(\mathbf{x}) \\
&= \mathbf{T1(t)}^T\mathbf{MT1}(\mathbf{x}) + \mathbf{T2(t)}^T\mathbf{NT1}(\mathbf{x}) \\
&\quad + \mathbf{T1(t)}^T\mathbf{PT2}(\mathbf{x}) + \mathbf{T2(t)}^T\mathbf{QT2}(\mathbf{x}) \qquad (7.9)
\end{aligned}$$

Substituting equations (7.3) and (7.9) in the last term of equation (7.1), we have

$$\begin{aligned}
&\lambda \int_a^b K(x, t)y(t)dt \\
&= \lambda \int_a^b [\mathbf{E}^T\mathbf{T1(t)} + \mathbf{F}^T\mathbf{T2(t)}] \times [\mathbf{T1(t)}^T\mathbf{MT1}(\mathbf{x}) \\
&\quad + \mathbf{T2(t)}^T\mathbf{NT1}(\mathbf{x}) + \mathbf{T1(t)}^T\mathbf{PT2}(\mathbf{x}) + \mathbf{T2(t)}^T\mathbf{QT2}(\mathbf{x})]dt
\end{aligned}$$
$$(7.10)$$

Using *Lemma 3* of Section 2.7, we write

$$\begin{aligned}
&= \lambda\mathbf{E}^T \int_a^b \{\mathbf{T1(t)T1(t)}^T dt\}\mathbf{MT1}(\mathbf{x}) \\
&\quad + \lambda\mathbf{E}^T \int_a^b \{\mathbf{T1(t)T1(t)}^T dt\}\mathbf{PT2}(\mathbf{x})
\end{aligned}$$

$$+ \lambda \mathbf{F}^{\mathrm{T}} \int_a^b \{\mathbf{T2}(t)\mathbf{T2}(t)^{\mathrm{T}} dt\} \mathbf{NT1}(\mathbf{x})$$

$$+ \lambda \mathbf{F}^{\mathrm{T}} \int_a^b \{\mathbf{T2}(t)\mathbf{T2}(t)^{\mathrm{T}} dt\} \mathbf{QT2}(\mathbf{x})$$

Using the results of Section 2.7.1, we write

$$\lambda \int_a^b K(x,t)y(t)dt = \lambda \mathbf{E}^{\mathrm{T}}\{\mathbf{\Phi 1}(\mathbf{b}) - \mathbf{\Phi 1}(\mathbf{a})\}\mathbf{MT1}(\mathbf{x})$$
$$+ \lambda \mathbf{E}^{\mathrm{T}}\{\mathbf{\Phi 1}(\mathbf{b}) - \mathbf{\Phi 1}(\mathbf{a})\}\mathbf{PT2}(\mathbf{x})$$
$$+ \lambda \mathbf{F}^{\mathrm{T}}\{\mathbf{\Phi 2}(\mathbf{b}) - \mathbf{\Phi 2}(\mathbf{a})\}\mathbf{NT1}(\mathbf{x})$$
$$+ \lambda \mathbf{F}^{\mathrm{T}}\{\mathbf{\Phi 2}(\mathbf{b}) - \mathbf{\Phi 2}(\mathbf{a})\}\mathbf{QT2}(\mathbf{x}) \quad (7.11)$$

Substituting equations (7.4), (7.5), and (7.11) in equation (7.1) we have

$$\mathbf{A}^{\mathrm{T}} * \mathbf{E}^{\mathrm{T}}\mathbf{T1}(\mathbf{x}) + \mathbf{B}^{\mathrm{T}} * \mathbf{F}^{\mathrm{T}}\mathbf{T2}(\mathbf{x})$$
$$= [\mathbf{G}^{\mathrm{T}}\mathbf{T1}(\mathbf{x}) + \mathbf{H}^{\mathrm{T}}\mathbf{T2}(\mathbf{x})] + \lambda \mathbf{E}^{\mathrm{T}}\{\mathbf{\Phi 1}(\mathbf{b}) - \mathbf{\Phi 1}(\mathbf{a})\}\mathbf{MT1}(\mathbf{x})$$
$$+ \lambda \mathbf{E}^{\mathrm{T}}\{\mathbf{\Phi 1}(\mathbf{b}) - \mathbf{\Phi 1}(\mathbf{a})\}\mathbf{PT2}(\mathbf{x})$$
$$+ \lambda \mathbf{F}^{\mathrm{T}}\{\mathbf{\Phi 2}(\mathbf{b}) - \mathbf{\Phi 2}(\mathbf{a})\}\mathbf{NT1}(\mathbf{x})$$
$$+ \lambda \mathbf{F}^{\mathrm{T}}\{\mathbf{\Phi 2}(\mathbf{b}) - \mathbf{\Phi 2}(\mathbf{a})\}\mathbf{QT2}(\mathbf{x}) \quad (7.12)$$

Now, equating the coefficients of $\mathbf{T1}(\mathbf{x})$ and $\mathbf{T2}(\mathbf{x})$ in equation (7.12), we get

$$\mathbf{A}^{\mathrm{T}} * \mathbf{E}^{\mathrm{T}} = \mathbf{G}^{\mathrm{T}} + \lambda \mathbf{E}^{\mathrm{T}}\{\mathbf{\Phi 1}(\mathbf{b}) - \mathbf{\Phi 1}(\mathbf{a})\}\mathbf{M}$$
$$+ \lambda \mathbf{F}^{\mathrm{T}}\{\mathbf{\Phi 2}(\mathbf{b}) - \mathbf{\Phi 2}(\mathbf{a})\}\mathbf{N}$$
$$\mathbf{B}^{\mathrm{T}} * \mathbf{F}^{\mathrm{T}} = \mathbf{H}^{\mathrm{T}} + \lambda \mathbf{E}^{\mathrm{T}}\{\mathbf{\Phi 1}(\mathbf{b}) - \mathbf{\Phi 1}(\mathbf{a})\}\mathbf{P}$$
$$+ \lambda \mathbf{F}^{\mathrm{T}}\{\mathbf{\Phi 2}(\mathbf{b}) - \mathbf{\Phi 2}(\mathbf{a})\}\mathbf{Q}$$

Or, using equations (2.39), (2.43), and (2.46). we have

$$\mathbf{E}^{\mathrm{T}}\mathbf{a}_{(m \times m)} = \mathbf{G}^{\mathrm{T}} + \lambda \mathbf{E}^{\mathrm{T}}\mathbf{\Phi 1}'\mathbf{M} + \lambda \mathbf{F}^{\mathrm{T}}\mathbf{\Phi 2}'\mathbf{N} \quad (7.13)$$
$$\mathbf{F}^{\mathrm{T}}\mathbf{b}_{(m \times m)} = \mathbf{H}^{\mathrm{T}} + \lambda \mathbf{E}^{\mathrm{T}}\mathbf{\Phi 1}'\mathbf{P} + \lambda \mathbf{F}^{\mathrm{T}}\mathbf{\Phi 2}'\mathbf{Q} \quad (7.14)$$

where, **a** and **b** are diagonal matrices formed by the elements of **A** and **B**.

Solving for $\mathbf{E}^T$ and $\mathbf{F}^T$ from the simultaneous vector-matrix equations, we get

$$\mathbf{E}^T = [\mathbf{G}^T + \mathbf{H}^T\mathbf{X_q}^{-1}\mathbf{Y}][\mathbf{X_m} - \mathbf{V}\mathbf{X_q}^{-1}\mathbf{Y}]^{-1} \tag{7.15}$$

$$\mathbf{F}^T = [\mathbf{H}^T + \mathbf{E}^T\mathbf{V}]\mathbf{X_q}^{-1} \tag{7.16}$$

where, $\mathbf{X_q} = \mathbf{b}_{(m\times m)} - \lambda\boldsymbol{\Phi}\mathbf{2}'\mathbf{Q}$, $\mathbf{Y} = \lambda\boldsymbol{\Phi}\mathbf{2}'\mathbf{N}$, $\mathbf{X_m} = \mathbf{a}_{(m\times m)} - \lambda\boldsymbol{\Phi}\mathbf{1}'\mathbf{M}$ and $\mathbf{V} = \lambda\boldsymbol{\Phi}\mathbf{1}'\mathbf{P}$.

7.1.1.1 Numerical examples

Example 1 [6]
Consider the integral equation

$$y(x) = \left(\frac{3x}{2} - \frac{1}{3}\right) + \int_0^1 (t - x)y(t)dt \tag{7.17}$$

The exact solution of equation (7.17) is $y(x) = x$.

Let us take $m = 4$ and $T = 1$. Then according to equation (7.4),

$$\left.\begin{array}{l} \mathbf{G}^T = \begin{bmatrix} -\frac{1}{3} & \frac{1}{24} & \frac{10}{24} & \frac{19}{24} \end{bmatrix} \\[2mm] \mathbf{H}^T = \begin{bmatrix} \frac{1}{24} & \frac{10}{24} & \frac{19}{24} & \frac{28}{24} \end{bmatrix} \end{array}\right\} \tag{7.18}$$

Following equations (7.6)–(7.8), we have

$$\mathbf{M} = \begin{bmatrix} 0 & -\frac{1}{4} & -\frac{2}{4} & -\frac{3}{4} \\ \frac{1}{4} & 0 & -\frac{1}{4} & -\frac{2}{4} \\ \frac{2}{4} & \frac{1}{4} & 0 & -\frac{1}{4} \\ \frac{3}{4} & \frac{2}{4} & \frac{1}{4} & 0 \end{bmatrix}, \quad \mathbf{N} = \begin{bmatrix} \frac{1}{4} & 0 & -\frac{1}{4} & -\frac{2}{4} \\ \frac{2}{4} & \frac{1}{4} & 0 & -\frac{1}{4} \\ \frac{3}{4} & \frac{2}{4} & \frac{1}{4} & 0 \\ \frac{4}{4} & \frac{3}{4} & \frac{2}{4} & \frac{1}{4} \end{bmatrix}$$

$$\mathbf{P} = \begin{bmatrix} -\frac{1}{4} & -\frac{2}{4} & -\frac{3}{4} & -\frac{4}{4} \\ 0 & -\frac{1}{4} & -\frac{2}{4} & -\frac{3}{4} \\ \frac{1}{4} & 0 & -\frac{1}{4} & -\frac{2}{4} \\ \frac{2}{4} & \frac{1}{4} & 0 & -\frac{1}{4} \end{bmatrix}, \quad \mathbf{Q} = \begin{bmatrix} 0 & -\frac{1}{4} & -\frac{2}{4} & -\frac{3}{4} \\ \frac{1}{4} & 0 & -\frac{1}{4} & -\frac{2}{4} \\ \frac{2}{4} & \frac{1}{4} & 0 & -\frac{1}{4} \\ \frac{3}{4} & \frac{2}{4} & \frac{1}{4} & 0 \end{bmatrix}$$

Since, $\alpha(x) = 1$, we have, $\mathbf{a}_{(m \times m)} = \mathbf{b}_{(m \times m)} = \mathbf{I}_{(m \times m)}$.

For this case, $\mathbf{\Phi 1}' = \mathbf{\Phi 2}' = h/2\mathbf{I}_{(m \times m)}$.

Since, $\lambda = 1$, using above values in equations (7.15) and (7.16) we get

$$\begin{aligned} \mathbf{E}^{\mathrm{T}} &= [0.00142857 \; 0.26190474 \; 0.50952379 \; 0.75714282] \\ \mathbf{F}^{\mathrm{T}} &= [0.26190474 \; 0.50952379 \; 0.75714282 \; 1.00476187] \end{aligned}$$
$$(7.19)$$

Direct expansion of $y(x) = x$ in TF domain gives the vectors $\mathbf{E}'^{\mathrm{T}}$ and $\mathbf{F}'^{\mathrm{T}}$ as

$$\begin{aligned} \mathbf{E}'^{\mathrm{T}} &= [0 \; 0.25 \; 0.5 \; 0.75] \\ \mathbf{F}'^{\mathrm{T}} &= [0.25 \; 0.5 \; 0.75 \; 1.0] \end{aligned} \qquad (7.20)$$

Comparing equations (7.19) and (7.20), it is seen that the TF domain results are pretty close to the exact solution. Also, since BPF based analysis provides a staircase solution, it is apparent that TF domain solution will incur much less ISE [9,10] than the corresponding BPF domain solution.

In fact, using computer program, the ISE for BPF analysis turns out to be 5.208331E–03, while the TF analysis introduces an ISE of 3.890784E–05. The ratio of ISEs is

$$\frac{\mathrm{ISE_{BPF}}}{\mathrm{ISE_{TF}}} = 133.86327$$

This is also seen from Fig. 7.1.

Now, we take $m = 10$ and solve the same example over the same time interval $[0, T)$ and assess the improvement of the computed results.

$$\begin{aligned} \mathbf{E}^{\mathrm{T}} = [&0.00230414 \quad 0.10215053 \quad 0.20199693 \\ &0.30184332 \quad 0.40168972 \quad 0.50153609 \\ &0.60138253 \quad 0.70122880 \quad 0.80107527 \\ &0.90092163] \end{aligned}$$

$$\begin{aligned} \mathbf{F}^{\mathrm{T}} = [&0.10215053 \quad 0.20199693 \quad 0.30184332 \\ &0.40168972 \quad 0.50153609 \quad 0.60138253 \\ &0.70122880 \quad 0.80107527 \quad 0.90092163 \\ &1.00076801] \end{aligned}$$

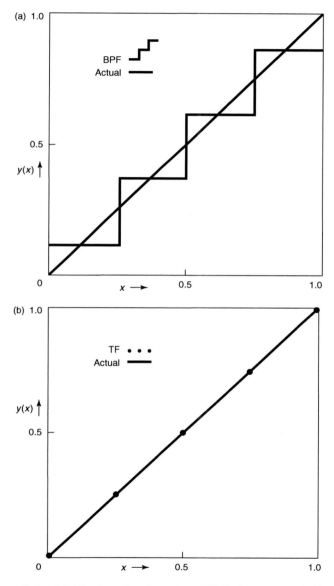

Figure 7.1. *(a) Block pulse function (BPF) domain and (b) triangular function (TF) domain solution of the integral equation of equation (7.17) for $T = 1$ s and $m = 4$.*

Direct expansion of $y(x) = x$ in TF domain gives

$$\mathbf{E}'^{\mathrm{T}} = [0\ 0.1\ 0.2\ 0.3\ 0.4\ 0.5\ 0.6\ 0.7\ 0.8\ 0.9]$$
$$\mathbf{F}'^{\mathrm{T}} = [0.1\ 0.2\ 0.3\ 0.4\ 0.5\ 0.6\ 0.7\ 0.8\ 0.9\ 1.0]$$

The results are compared in Fig. 7.2.

Example 2 [1]

Consider the integral equation

$$y(x) = \left[\sin(x) - \frac{x}{4}\right] + \frac{1}{4}\int_0^{\pi/2} txy(t)\mathrm{d}t \qquad (7.21)$$

The exact solution of equation (7.21) is $y(x) = \sin x$.
For $m = 10$ and $T = \frac{\pi}{2}$ s

$$\mathbf{E}^{\mathrm{T}} = [0\ \ 0.15655392\ \ 0.30925592\ \ 0.45434891$$
$$0.58826320\ \ 0.70770427\ \ 0.80973408\ \ 0.89184322$$
$$0.95201299\ \ 0.98876434]$$

$$\mathbf{F}^{\mathrm{T}} = [0.15655392\ \ 0.30925592\ \ 0.45434891\ \ 0.58826320$$
$$0.70770427\ \ 0.80973408\ \ 0.89184322\ \ 0.95201299$$
$$0.98876434\ \ 1.00119595]$$

Direct expansion of $y(x) = \sin x$ in triangular function domain yields.

$$\mathbf{E}'^{\mathrm{T}} = [0\ \ 0.15643434\ \ 0.30901676\ \ 0.45399016$$
$$0.58778488\ \ 0.70710635\ \ 0.80901658$$
$$0.89100617\ \ 0.95105630\ \ 0.98768812]$$

$$\mathbf{F}'^{\mathrm{T}} = [0.15643434\ \ 0.30901676\ \ 0.45399016\ \ 0.58778488$$
$$0.70710635\ \ 0.80901658\ \ 0.89100617\ \ 0.95105630$$
$$0.98768812\ \ 1.00000011]$$

Figure 7.3(a) shows the results obtained via equations (7.15) and (7.16), and direct expansion. It is seen that they are in good corroboration. Figure 7.3(b) shows the equivalent BPF domain solution.

Figure 7.4 compares the TF domain solution with BPF domain solution of the same problem for $m = 4$ and $T = \frac{\pi}{2}$ s.

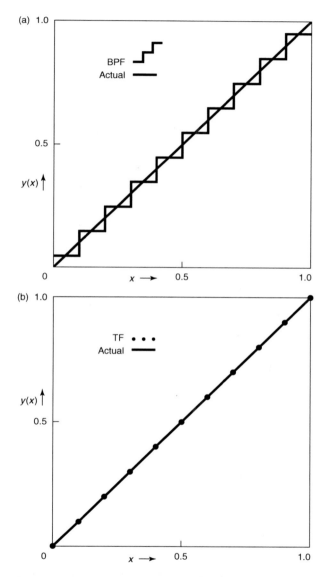

Figure 7.2. *(a) Block pulse function (BPF) domain and (b) triangular function (TF) domain solution of the integral equation of equation (7.17) for $T = 1$ s and $m = 10$.*

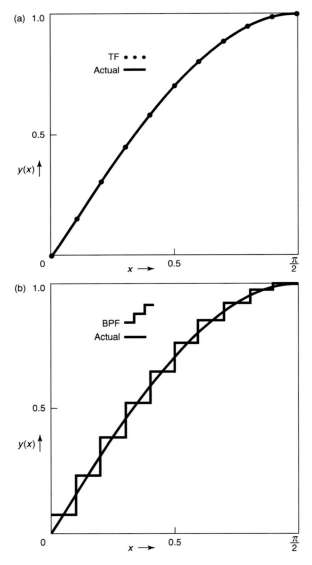

Figure 7.3. *(a) TF domain solution of equation (7.21) for $T = \frac{\pi}{2}$ s and $m = 10$, compared with the actual solution. (b) Block pulse function (BPF) domain solution of equation (7.21) for $T = \frac{\pi}{2}$ s and $m = 10$, compared with the exact solution.*

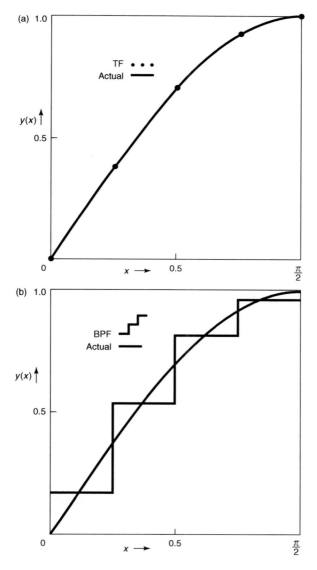

Figure 7.4. *(a) TF domain solution of equation (7.21) for $T = \frac{\pi}{2}s$ and $m = 4$, compared with the actual solution. (b) BPF domain solution of equation (7.21) for $T = \frac{\pi}{2}s$ and $m = 4$, compared with the actual solution.*

Example 3 [1]

Let us consider the equation

$$y(x) = \left[\exp(x) - \exp(1)/2 + \frac{1}{2}\right] + \frac{1}{2}\int_0^1 y(t)dt \qquad (7.22)$$

The exact solution of equation (7.22) is $y(x) = \exp(x)$.

For $m = 10$ and $T = 1$ s

$$\mathbf{E}^T = [1.00143174 \quad 1.10660270 \quad 1.22283450$$
$$1.35129050 \quad 1.49325636 \quad 1.65015296$$
$$1.82355050 \quad 2.01518444 \quad 2.22697261$$
$$2.46103481]$$

$$\mathbf{F}^T = [1.10660270 \quad 1.22283450 \quad 1.35129050$$
$$1.49325636 \quad 1.65015296 \quad 1.82355050$$
$$2.01518444 \quad 2.22697261 \quad 2.46103481$$
$$2.71971348]$$

Direct expansion of $y(x) = \exp(x)$ in TF domain yields

$$\mathbf{E}'^T = [1 \quad 1.10517096 \quad 1.22140276 \quad 1.34985876$$
$$1.49182462 \quad 1.64872121 \quad 1.82211875$$
$$2.01375269 \quad 2.22554087 \quad 2.45960307]$$

$$\mathbf{F}'^T = [1.10517096 \quad 1.22140276 \quad 1.34985876$$
$$1.49182462 \quad 1.64872121 \quad 1.82211875$$
$$2.01375269 \quad 2.22554087 \quad 2.45960307$$
$$2.71828174]$$

Figure 7.5 compares the two sets of results and they are very close.

7.1.2 Volterra integral equation of the second kind

Such an equation is given by [1]

$$\alpha(x)y(x) = f(x) + \lambda \int_a^x K(x,t)y(t)dt \qquad (7.23)$$

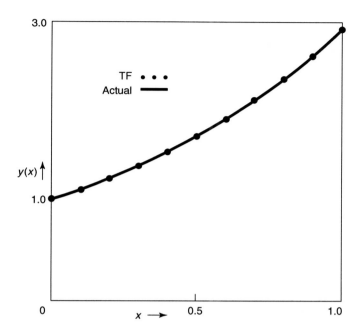

Figure 7.5. *TF domain solution of equation (7.22) for T = 1 s and m = 10, compared with the actual solution.*

where, as before, $\alpha(x), f(x)$, and the Kernel $K(x, t)$ are given. And a and λ are constants.

We proceed in a manner similar to Section 7.1.1 and expand $\alpha(x), y(x), f(x)$, and $K(x, t)$ into respective m-set TF series over the relevant interval.

Thus, in line with equation (7.10), we can write

$$\lambda \int_a^x K(x, t)y(t)\mathrm{d}t = \lambda \int_a^x [\mathbf{E}^{\mathrm{T}}\mathbf{T1}(\mathbf{t}) + \mathbf{F}^{\mathrm{T}}\mathbf{T2}(\mathbf{t})][\mathbf{T1}(\mathbf{t})^{\mathrm{T}}\mathbf{MT1}(\mathbf{x})$$
$$+ \mathbf{T2}(\mathbf{t})^{\mathrm{T}}\mathbf{NT1}(x) + \mathbf{T1}(\mathbf{t})^{\mathrm{T}}\mathbf{PT2}(x)$$
$$+ \mathbf{T2}(\mathbf{t})^{\mathrm{T}}\mathbf{QT2}(\mathbf{x})]\mathrm{d}t \qquad (7.24)$$
$$= \lambda\mathbf{E}^{\mathrm{T}} \int_a^x \{\mathbf{T1}(\mathbf{t})\mathbf{T1}(\mathbf{t})^{\mathrm{T}}\mathrm{d}t\}\mathbf{MT1}(\mathbf{x})$$
$$+ \lambda\mathbf{E}^{\mathrm{T}} \int_a^x \{\mathbf{T1}(\mathbf{t})\mathbf{T1}(\mathbf{t})^{\mathrm{T}}\mathrm{d}t\}\mathbf{PT2}(x)$$

$$+ \lambda \mathbf{F}^T \int_a^x \{\mathbf{T2}(t)\mathbf{T2}(t)^T dt\}\mathbf{NT1}(x)$$

$$+ \lambda \mathbf{F}^T \int_a^x \{\mathbf{T2}(t)\mathbf{T2}(t)^T dt\}\mathbf{QT2}(x)$$

$$(7.25)$$

All the four integrals of equation (7.25) are of the same pattern. Hence, we evaluate the first integral, say I_1, in the following. Dropping the constant term before the integral sign, we can write

$$I_1 = \int_a^x \{\mathbf{T1}(t)\mathbf{T1}(t)^T dt\}\mathbf{MT1}(x)$$

$$= \int_0^x \{\mathbf{T1}(t)\mathbf{T1}(t)^T dt\}\mathbf{MT1}(x) - \int_0^a \{\mathbf{T1}(t)\mathbf{T1}(t)^T dt\}\mathbf{MT1}(x)$$

Using the result of Section 2.7.1, we write

$$= \int_0^x \{\mathrm{diag}[\mathrm{T1}_0 \; \mathrm{T1}_1 \; \mathrm{T1}_2 \quad \ldots \; \mathrm{T1}_{(m-1)}]dt\}\mathbf{MT1}(x)$$

$$- \int_0^a \{\mathrm{diag}[\mathrm{T1}_0 \; \mathrm{T1}_1 \; \mathrm{T1}_2 \quad \ldots \; \mathrm{T1}_{(m-1)}]dt\}\mathbf{MT1}(x) \quad (7.26)$$

We know that

$$\int_0^x \mathrm{T1}_i(t)dt = \mathbf{h1}_i^T \mathbf{T1}_{(m)}(x) + \mathbf{h2}_i^T \mathbf{T2}_{(m)}(x) \qquad (7.27)$$

for $i = 0, 1, 2, \ldots, (m - 1)$, where $\mathbf{h1}_i^T$ is the ith row of the integration operational matrix $\mathbf{P1}_{(m \times m)}$ and $\mathbf{h2}_i^T$ is the ith row of the integration operational matrix $\mathbf{P2}_{(m \times m)}$.

Substituting equation (7.27) in equation (7.26), we have

$$I_1 = \begin{bmatrix} \mathbf{h1}_0^T\mathbf{T1}(x) + \mathbf{h2}_0^T\mathbf{T2}(x) & 0 & \ldots & \ldots & \ldots & 0 \\ 0 & \mathbf{h1}_1^T\mathbf{T1}(x) + \mathbf{h2}_1^T\mathbf{T2}(x) & 0 & \ldots & \ldots & 0 \\ \ldots & \ldots & \ldots & \ldots & \ldots & \ldots & \ldots & \ldots \\ \ldots & \ldots & \ldots & \ldots & \ldots & \ldots & \ldots & \ldots \\ \ldots & \ldots & \ldots & \ldots & \ldots & \ldots & \ldots & \ldots \\ 0 & 0 & \ldots & 0 & \mathbf{h1}_{(m-1)}^T\mathbf{T1}(x) + \mathbf{h2}_{(m-1)}^T\mathbf{T2}(x) \end{bmatrix}$$

$$\times \begin{bmatrix} \mathbf{m0r}^T\mathbf{T1(x)} \\ \mathbf{m1r}^T\mathbf{T1(x)} \\ \mathbf{m2r}^T\mathbf{T1(x)} \\ \dots \\ \dots \\ \mathbf{m(m-1)r}^T\mathbf{T1(x)} \end{bmatrix} - [\mathbf{\Phi1(a)} - \mathbf{\Phi1(0)}]\mathbf{MT1(x)} \tag{7.28}$$

where, $\mathbf{M} = [\mathbf{m0r} \ \mathbf{m1r} \ \mathbf{m2r} \ \dots \ \mathbf{mir} \ \dots \ \mathbf{m(m-1)r}]^T$ and $\mathbf{mir}^T$ is the ith row of the $(m \times m)$ matrix $\mathbf{M}$. Also, in equation (7.28), we have dropped (m) in the subscript of $\mathbf{T1_{(m)}(x)}$ for simplicity.

Since $\mathbf{\Phi1(0)} = 0$ and using the results of *Lemmas 1* to *3* and *Theorem 1 of Section 2.7*, we have

$$I_1 = \begin{bmatrix} \mathbf{h1}_0^T\mathbf{M0r_{(m \times m)}T1(x)} \\ \mathbf{h1}_1^T\mathbf{M1r_{(m \times m)}T1(x)} \\ \dots \\ \dots \\ \mathbf{h1}_{(m-1)}^T\mathbf{M(m-1)r_{(m \times m)}T1(x)} \end{bmatrix}_{(m \times m)} - \mathbf{\Phi1(a)MT1(x)} \tag{7.29}$$

where, $\mathbf{Mir_{(m \times m)}} = \text{diag}[mi0 \ mi1 \ mi2 \ \dots \ \dots \ mi(m-1)]$, that is $\mathbf{Mir_{(m \times m)}}$ is a diagonal matrix formed by the elements of the ith row of the martix $\mathbf{M}$.

Now let

$$\begin{bmatrix} \mathbf{h1}_0^T\mathbf{M0r_{(m \times m)}T1(x)} \\ \mathbf{h1}_1^T\mathbf{M1r_{(m \times m)}T1(x)} \\ \dots \\ \dots \\ \mathbf{h1}_{(m-1)}^T\mathbf{M(m-1)r_{(m \times m)}T1(x)} \end{bmatrix} = \mathbf{K1 \ T1(x)} \tag{7.30}$$

Then equation (7.29) may be written as

$$I_1 = [\mathbf{K1} - \mathbf{\Phi1(a)M}]\mathbf{T1(x)} \tag{7.31}$$

Similar expressions may be obtained for other three integrals of equation (7.25). These are, say,

$$
I_2 = \begin{bmatrix} \mathbf{h2}_0^\mathsf{T}\mathbf{P0r}_{(m \times m)}\mathbf{T2}(\mathbf{x}) \\ \mathbf{h2}_1^\mathsf{T}\mathbf{P1r}_{(m \times m)}\mathbf{T2}(\mathbf{x}) \\ \cdots \\ \cdots \\ \mathbf{h2}_{(m-1)}^\mathsf{T}\mathbf{P(m-1)r}_{(m \times m)}\mathbf{T2}(\mathbf{x}) \end{bmatrix}_{(m \times m)} - \mathbf{\Phi1}(\mathbf{a})\mathbf{PT2}(\mathbf{x})
$$

$$
= [\mathbf{K2} - \mathbf{\Phi1}(\mathbf{a})\mathbf{P}]\mathbf{T2}(\mathbf{x}) \tag{7.32}
$$

$$
I_3 = \begin{bmatrix} \mathbf{h1}_0^\mathsf{T}\mathbf{N0r}_{(m \times m)}\mathbf{T1}(\mathbf{x}) \\ \mathbf{h1}_1^\mathsf{T}\mathbf{N1r}_{(m \times m)}\mathbf{T1}(\mathbf{x}) \\ \cdots \\ \cdots \\ \mathbf{h1}_{(m-1)}^\mathsf{T}\mathbf{N(m-1)r}_{(m \times m)}\mathbf{T1}(\mathbf{x}) \end{bmatrix}_{(m \times m)} - \mathbf{\Phi2}(\mathbf{a})\mathbf{NT1}(\mathbf{x})
$$

$$
= [\mathbf{K3} - \mathbf{\Phi2}(\mathbf{a})\mathbf{N}]\mathbf{T1}(\mathbf{x}) \tag{7.33}
$$

$$
I_4 = \begin{bmatrix} \mathbf{h2}_0^\mathsf{T}\mathbf{Q0r}_{(m \times m)}\mathbf{T2}(\mathbf{x}) \\ \mathbf{h2}_1^\mathsf{T}\mathbf{Q1r}_{(m \times m)}\mathbf{T2}(\mathbf{x}) \\ \cdots \\ \cdots \\ \mathbf{h2}_{(m-1)}^\mathsf{T}\mathbf{Q(m-1)r}_{(m \times m)}\mathbf{T2}(\mathbf{x}) \end{bmatrix}_{(m \times m)} - \mathbf{\Phi2}(\mathbf{a})\mathbf{QT2}(\mathbf{x})
$$

$$
= [\mathbf{K4} - \mathbf{\Phi2}(\mathbf{a})\mathbf{Q}]\mathbf{T2}(\mathbf{x}) \tag{7.34}
$$

Substituting equations (7.31)–(7.34) in equation (7.25), we have

$$
\lambda \int_a^x K(x,t)y(t)dt = \lambda\mathbf{E}^\mathsf{T}[\mathbf{K1} - \mathbf{\Phi1}(\mathbf{a})\mathbf{M}]\mathbf{T1}(\mathbf{x})
$$
$$
+ \lambda\mathbf{E}^\mathsf{T}[\mathbf{K2} - \mathbf{\Phi1}(\mathbf{a})\mathbf{P}]\mathbf{T2}(\mathbf{x})
$$
$$
+ \lambda\mathbf{F}^\mathsf{T}[\mathbf{K3} - \mathbf{\Phi2}(\mathbf{a})\mathbf{N}]\mathbf{T1}(\mathbf{x})
$$
$$
+ \lambda\mathbf{F}^\mathsf{T}[\mathbf{K4} - \mathbf{\Phi2}(\mathbf{a})\mathbf{Q}]\mathbf{T2}(\mathbf{x}) \tag{7.35}
$$

Substituting RHS of equation (7.35) in equation (7.23) and using equations (7.2)–(7.5), we get

$$\mathbf{E}^T \mathbf{T1}(\mathbf{x}) + \mathbf{F}^T \mathbf{T2}(\mathbf{x}) = \mathbf{G}^T \mathbf{T1}(\mathbf{x}) + \mathbf{H}^T \mathbf{T2}(\mathbf{x})$$
$$+ \lambda \mathbf{E}^T [\mathbf{K1} - \mathbf{\Phi1}(\mathbf{a})\mathbf{M}] \mathbf{T1}(\mathbf{x})$$
$$+ \lambda \mathbf{E}^T [\mathbf{K2} - \mathbf{\Phi1}(\mathbf{a})\mathbf{P}] \mathbf{T2}(\mathbf{x})$$
$$+ \lambda \mathbf{F}^T [\mathbf{K3} - \mathbf{\Phi2}(\mathbf{a})\mathbf{N}] \mathbf{T1}(\mathbf{x})$$
$$+ \lambda \mathbf{F}^T [\mathbf{K4} - \mathbf{\Phi2}(\mathbf{a})\mathbf{Q}] \mathbf{T2}(\mathbf{x}) \quad (7.36)$$

Comparing the coefficients of $\mathbf{T1}(\mathbf{x})$ and $\mathbf{T2}(\mathbf{x})$ on both sides separately, we have

$$\mathbf{E}^T = \mathbf{G}^T + \lambda \mathbf{E}^T \mathbf{K1} + \lambda \mathbf{F}^T \mathbf{K3} - \lambda \mathbf{E}^T \mathbf{\Phi1}(\mathbf{a})\mathbf{M} - \lambda \mathbf{F}^T \mathbf{\Phi2}(\mathbf{a})\mathbf{N} \quad (7.37)$$

$$\mathbf{F}^T = \mathbf{H}^T + \lambda \mathbf{E}^T \mathbf{K2} + \lambda \mathbf{F}^T \mathbf{K4} - \lambda \mathbf{E}^T \mathbf{\Phi1}(\mathbf{a})\mathbf{P} - \lambda \mathbf{F}^T \mathbf{\Phi2}(\mathbf{a})\mathbf{Q} \quad (7.38)$$

or,

$$\mathbf{E}^T [\mathbf{I} - \lambda \mathbf{K1} + \lambda \mathbf{\Phi1}(\mathbf{a})\mathbf{M}] - \lambda \mathbf{F}^T [\mathbf{K3} - \mathbf{\Phi2}(\mathbf{a})\mathbf{N}] = \mathbf{G}^T \quad (7.39)$$

$$\mathbf{F}^T [\mathbf{I} - \lambda \mathbf{K4} + \lambda \mathbf{\Phi2}(\mathbf{a})\mathbf{Q}] - \lambda \mathbf{E}^T [\mathbf{K2} - \mathbf{\Phi1}(\mathbf{a})\mathbf{P}] = \mathbf{H}^T \quad (7.40)$$

Writing

$$\left. \begin{array}{l} \lambda[\mathbf{K1} - \mathbf{\Phi1}(\mathbf{a})\mathbf{M}] = \mathbf{U} \\ \lambda[\mathbf{K3} - \mathbf{\Phi2}(\mathbf{a})\mathbf{N}] = \mathbf{W} \\ \lambda[\mathbf{K4} - \mathbf{\Phi2}(\mathbf{a})\mathbf{Q}] = \mathbf{Z} \\ \lambda[\mathbf{K2} - \mathbf{\Phi1}(\mathbf{a})\mathbf{P}] = \mathbf{V} \end{array} \right\} \quad (7.41)$$

we have

$$\mathbf{E}^T(\mathbf{I} - \mathbf{U}) - \mathbf{F}^T \mathbf{W} = \mathbf{G}^T \quad (7.42)$$

$$\mathbf{F}^T(\mathbf{I} - \mathbf{Z}) - \mathbf{E}^T \mathbf{V} = \mathbf{H}^T \quad (7.43)$$

From equation (7.43),

$$\mathbf{F}^T = [\mathbf{H}^T + \mathbf{E}^T \mathbf{V}](\mathbf{I} - \mathbf{Z})^{-1} \quad (7.44)$$

Post-multiplying both sides of equation (7.43) by $(\mathbf{I} - \mathbf{Z})^{-1}\mathbf{W}$, we have

$$\mathbf{F}^T \mathbf{W} - \mathbf{E}^T \mathbf{V}(\mathbf{I} - \mathbf{Z})^{-1}\mathbf{W} = \mathbf{H}^T(\mathbf{I} - \mathbf{Z})^{-1}\mathbf{W}$$

Substituting $\mathbf{F}^T\mathbf{W}$ in equation (7.42) and solving for $\mathbf{E}^T$, we have

$$\mathbf{E}^T = [\mathbf{G}^T + \mathbf{H}^T(\mathbf{I} - \mathbf{Z})^{-1}\mathbf{W}][(\mathbf{I} - \mathbf{U}) - \mathbf{V}(\mathbf{I} - \mathbf{Z})^{-1}\mathbf{W}]^{-1} \quad (7.45)$$

Equations (7.44) and (7.45) are used to solve for the vectors $\mathbf{E}$ and $\mathbf{F}$.

7.1.2.1 Numerical examples

Example 1 [6]

Consider the integral equation

$$y(x) = 3x + \int_0^x \sinh(x - t)y(t)dt \quad (7.46)$$

The exact solution of equation (7.46) is $y(x) = 3x/2 + (3\sqrt{2}/4)\sinh(x\sqrt{2})$. Let us take $m = 4$ and $T = 1\,s$. Then according to equation (7.4),

$$\left.\begin{array}{l}\mathbf{G}^T = [0 \quad 3/4 \quad 6/4 \quad 9/4] \\ \mathbf{H}^T = [3/4 \quad 6/4 \quad 9/4 \quad 12/4]\end{array}\right\} \quad (7.47)$$

Following equation (7.41), we have

$$\mathbf{U} = \frac{1}{8}\begin{bmatrix} 0 & 0.2526 & 0.5211 & 0.8223 \\ 0 & 0 & 0.2526 & 0.5211 \\ 0 & 0 & 0 & 0.2526 \\ 0 & 0 & 0 & 0 \end{bmatrix}$$

$$\mathbf{V} = \frac{1}{8}\begin{bmatrix} 0.2526 & 0.5211 & 0.8223 & 1.175 \\ 0 & 0.2526 & 0.5211 & 0.8223 \\ 0 & 0 & 0.2526 & 0.5211 \\ 0 & 0 & 0 & 0.2526 \end{bmatrix}$$

$$\mathbf{W} = \frac{1}{8}\begin{bmatrix} 0 & 0 & 0.2526 & 0.5211 \\ 0 & 0 & 0 & 0.2526 \\ 0 & 0 & 0 & 0 \\ 0 & 0 & 0 & 0 \end{bmatrix}$$

$$\mathbf{Z} = \frac{1}{8}\begin{bmatrix} 0 & 0.2526 & 0.5211 & 0.8223 \\ 0 & 0 & 0.2526 & 0.5211 \\ 0 & 0 & 0 & 0.2526 \\ 0 & 0 & 0 & 0 \end{bmatrix}$$

Since, $\alpha(x) = 1$, we have, $\mathbf{a}_{(m \times m)} = \mathbf{b}_{(m \times m)} = \mathbf{I}_{(m \times m)}$.
For this case, $\mathbf{\Phi 1}' = \mathbf{\Phi 2}' = (h/2)\mathbf{I}_{(m \times m)}$.
Since, $\lambda = 1$, using above values in equations (7.44) and (7.45) we get

$$\left. \begin{array}{l} \mathbf{E}^T = [0 \ 0.75 \ 1.54736481 \ 2.44542621] \\ \mathbf{F}^T = [0.75 \ 1.54736481 \ 2.44542621 \ 3.51020169] \end{array} \right\}$$

$$(7.48)$$

Direct expansion of $y(x) = 3x/2 + (3\sqrt{2}/4)\sinh(x\sqrt{2})$ in TF domain gives the vectors $\mathbf{E}'^T$ and $\mathbf{F}'^T$ as

$$\left. \begin{array}{l} \mathbf{E}'^T = [0 \ 0.75786143 \ 1.56408119 \ 2.47312545] \\ \mathbf{F}'^T = [0.75786143 \ 1.56408119 \ 2.47312545 \ 3.55244803] \end{array} \right\}$$

$$(7.49)$$

Comparing equations (7.48) and (7.49), it is seen that the TF domain results are pretty close to the exact solution. This is seen from Fig. 7.6.

Now, we take $m = 10$ and solve the same example over the same time interval to obtain

$$\begin{array}{ll} \mathbf{E}^T = [0 & 0.30000001 \quad 0.60300502 \quad 0.91208022 \\ & 1.23041232 \quad 1.56137339 \quad 1.90858860 \\ & 2.27600804 \quad 2.66798745 \quad 3.08937311] \\ \mathbf{F}^T = [0.30000001 & 0.60300502 \quad 0.91208022 \\ & 1.23041232 \quad 1.56137339 \quad 1.90858860 \\ & 2.27600804 \quad 2.66798745 \quad 3.08937311 \\ & 3.54560091] \end{array}$$

Direct expansion of $y(x) = 3x/2 + (3\sqrt{2}/4)\sinh(x\sqrt{2})$ in TF domain yields

$$\begin{array}{ll} \mathbf{E}'^T = [0 & 0.30050054 \quad 0.60401600 \quad 0.91362202 \\ & 1.23251581 \quad 1.56408119 \quad 1.91195535 \\ & 2.28010201 \quad 2.67289209 \quad 3.09518909] \end{array}$$

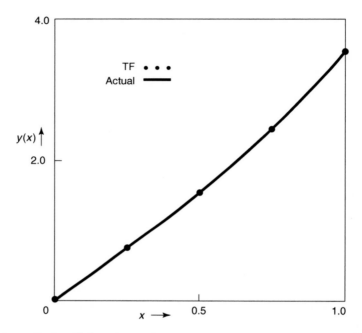

Figure 7.6. *TF domain solution of equation (7.46) for T = 1 s and m = 4, compared with the actual solution.*

$$\mathbf{F}'^{T} = [0.30050054 \quad 0.60401600 \quad 0.91362202$$
$$1.23251581 \quad 1.56408119 \quad 1.91195535$$
$$2.28010201 \quad 2.67289209 \quad 3.09518909$$
$$3.55244803]$$

Two sets of results are compared in Fig. 7.7 and is in good corroboration.

Example 2 [1]

Let us consider the equation

$$y(x) = \cos x - x - 2 + \int_0^x (t - x)y(t)\mathrm{d}t \qquad (7.50)$$

The exact solution of equation (7.50) is

$$y(x) = -[(x/2)\sin x + \sin x + \cos x]$$

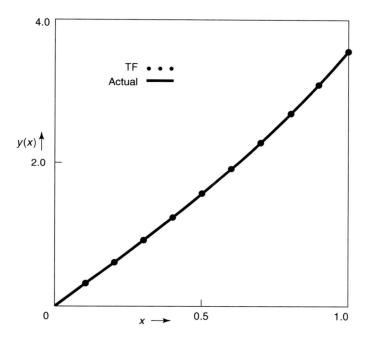

Figure 7.7. *TF domain solution of equation (7.46) for T = 1 s and m = 10, compared with the actual solution.*

For $m = 10$ and $T = 1$ s, since, $\lambda = 1$, using equations (7.44) and (7.45) we get

$$
\begin{aligned}
\mathbf{E}^{\mathrm{T}} = [-1 \quad &-1.09999596 \quad -1.19893343 \quad -1.29567412 \\
&-1.38900376 \quad -1.47764608 \quad -1.56028049 \\
&-1.63555855 \quad -1.70212323 \quad -1.75862781]
\end{aligned}
$$

$$
\begin{aligned}
\mathbf{F}^{\mathrm{T}} = [&-1.09999596 \quad -1.19893343 \quad -1.29567412 \\
&-1.38900376 \quad -1.47764608 \quad -1.56028049 \\
&-1.63555855 \quad -1.70212323 \quad -1.75862781 \\
&-1.80375697]
\end{aligned}
$$

Direct expansion of $y(x) = -[(x/2)\sin x + \sin x + \cos x]$ in TF domain yields

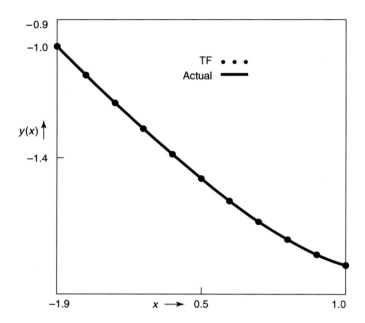

Figure 7.8. *TF domain solution of equation (7.50) for T = 1 s and m = 10, compared with the exact solution.*

$$\mathbf{E}'^{T} = [-1 \quad -1.09982919 \quad -1.19860291 \quad -1.29518485$$
$$-1.38836300 \quad -1.47686445 \quad -1.55937075$$
$$-1.63453614 \quad -1.70100522 \quad -1.75743412]$$
$$\mathbf{F}'^{T} = [-1.09982919 \quad -1.19860291 \quad -1.29518485$$
$$-1.38836300 \quad -1.47686445 \quad -1.55937075$$
$$-1.63453614 \quad -1.70100522 \quad -1.75743412$$
$$-1.80250883]$$

Figure 7.8 compares the two sets of results and they are very close.

Example 3 [1]

Consider the integral equation

$$y(x) = x + \int_{0}^{x} (t - x)y(t)\mathrm{d}t \qquad (7.51)$$

The exact solution of (7.51) is $y(x) = \sin x$.

For $m = 10$ and $T = 1$ s, since $\lambda = 1$, using equations (7.44) and (7.45) we get

$$\mathbf{E}^{\mathrm{T}} = [0 \quad 0.10000000 \quad 1.19900000 \quad 0.29601001$$
$$0.39005990 \quad 0.48020920 \quad 0.56555642$$
$$0.64524804 \quad 0.71848722 \quad 0.784541534]$$
$$\mathbf{F}^{\mathrm{T}} = [0.10000000 \quad 1.19900000 \quad 0.29601001$$
$$0.39005990 \quad 0.48020920 \quad 0.56555642$$
$$0.64524804 \quad 0.71848722 \quad 0.784541534$$
$$0.84275038]$$

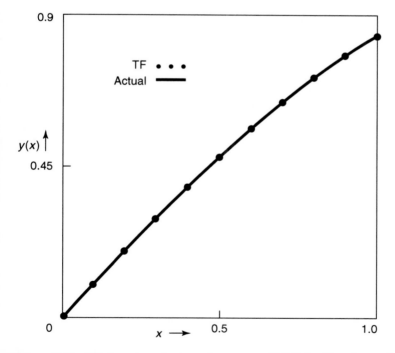

Figure 7.9. *TF domain solution of equation (7.51) for $T = 1$ s and $m = 10$, compared with the actual solution.*

Direct expansion of $y(x) = \sin x$ in TF domain yields

$$\mathbf{E}'^{T} = [0 \quad 0.09983341 \quad 0.19866932 \quad 0.29552018$$
$$0.38941836 \quad 0.47942551 \quad 0.56464248$$
$$0.64421766 \quad 0.71735614 \quad 0.78332698]$$
$$\mathbf{F}'^{T} = [0.09983341 \quad 0.19866932 \quad 0.29552018$$
$$0.38941836 \quad 0.47942551 \quad 0.56464248$$
$$0.64421766 \quad 0.71735614 \quad 0.78332698$$
$$0.84147101]$$

Figure 7.9 compares the two sets of results and they are very close.

7.2 Conclusion

Both Fredholm integral equation and Volterra integral equation are solved, using orthogonal triangular function set. The estimated ISE is found to be much less than the alternative staircase solution obtained from BPF domain analysis. From Figs 7.1 to 7.4, it is evident that TF domain analysis is much more effective than BPF domain analysis. Also, the effectiveness increases with increase of complexity of the functions to be handled.

References

1. Lovitt, W. V., *Linear integral equation*, Dover Publications, New York, 1924.
2. Deb, A., Sarkar, G. and Sen, S. K., Block pulse functions, the most fundamental of all piecewise constant basis functions, Int. J. Syst. Sci., vol. **25**, no. 2, pp. 351–363, 1994.
3. Jiang, J. H. and Schaufelberger, W., *Block pulse functions and their applications in control systems*, LNCIS-179, Springer-Verlag, Berlin, 1992.
4. Tzafestas, S. G. (Ed), *Walsh functions in signal and systems analysis and design*, Van-Nostrand Reinhold Co., New York, 1985.

5. Chen, C. F., Tsay, Y. T. and Wu, T. T., Walsh operational matrices for fractional calculus and their application to distributed systems, J. Franklin Instt., vol. **303**, no. 3, pp. 267–284, 1977.
6. Kung, F. C. and Chen, S. Y., Solution of integral equation using a set of block pulse functions, J. Franklin Instt., vol. **306**, no. 4, pp. 283–291, 1978.
7. Wang, Chi-Hsu and Shih, Yen-Ping, Explicit solutions of integral equations via block pulse functions, Int. J. Syst. Sci., vol. **13**, no. 7, pp. 773–782, 1982.
8. Chang, Rong-Yeu, Yang, Shwu-Yien and Wang, Maw-Ling, Solution of integral equations via generalized orthogonal polymials, Int. J. Syst. Sci., vol. **18**, no. 3, pp. 553–568, 1987.
9. Rao, G. P. and Srinivasan, T., Analysis and synthesis of dynamic systems containing time delays via block pulse functions, Proc. IEE, vol. **125**, no. 9, pp. 1064–1068, 1978.
10. Deb, Anish, Sarkar, Gautam and Sen, Sunit K., Linearly pulse-width modulated block pulse functions and their application to linear SISO feedback control system identification, IEE Proc. Control Theory and Appl., vol. **142**, no. 1, pp. 44–50, 1995.

Chapter 8

Microprocessor Based Simulation of Control Systems Using Orthogonal Functions

Control systems can be simulated more easily via orthogonal functions where the expansion coefficients are derived from samples of the related functions. With block pulse functions [1–3], computation of coefficients involves evaluation of integrations. Number of such integrations is equal to the number of component block pulse functions. However, with triangular functions [4], sample-and-hold functions [5], and the well-known Dirac delta functions [6], these integrations are avoided altogether, because, as per definition, expansion with such functions determines the coefficients from the samples of the concerned function.

The following experimental work presents microprocessor-based simulation of discrete time as well as sample-and-hold systems via sample-and-hold function set and Dirac delta function set.

A few open-loop and closed-loop sampled-data systems—with or without hold device—have been studied and the results obtained are compared with exact solutions.

8.1 Review of Delta Function and Sample-and-Hold Function Operational Technique

8.1.1 Delta function domain operational technique [6]

If a square integrable time function $f(t)$ is fed to a sampling device, the device modulates $f(t)$ to $f^*(t)$ which is obtained as the output. Hence

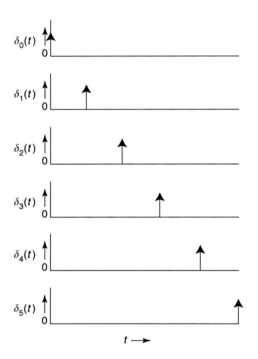

Figure 8.1. *A set of delta functions.*

$$f^*(t) = \sum_{i=0}^{m-1} f_i \delta_i(t) = \begin{bmatrix} f_0 & f_1 & f_2 & \cdots & f_i & \cdots & f_{(m-1)} \end{bmatrix} \mathbf{\Delta_{(m)}(t)}$$

$$\triangleq \mathbf{F_{(m)}} \mathbf{\Delta_{(m)}(t)} \qquad (8.1)$$

where we have chosen an m-set Dirac delta functions given by
$\mathbf{\Delta_{(m)}(t)} \triangleq [\delta_0(t) \quad \delta_1(t) \quad \cdots \quad \delta_i(t) \quad \cdots \quad \delta_{(m-1)}(t)]^{\mathsf{T}}$,
$[\ldots]^{\mathsf{T}}$ denotes transpose.

The component delta functions are delayed delta functions,
given by

$$\delta_i(t) \triangleq \delta(t - ih)$$

where, $i = 0, 1, 2, \ldots, (m - 1)$ and $h = T/m$, T is the time
period under consideration and m is the number of component
functions.

Figure 8.1 shows a set of delta functions.

The coefficients f_i's of equation (8.1) are given by

$$f_i = \int_0^T f(t)\delta(t - ih)dt, \quad ih \leq T \qquad (8.2)$$

The operational matrix for integration of order m in the delta function (DF) domain is the upper triangular matrix $\mathbf{H1}_{(m)}$ expressed as

$$\mathbf{H1}_{(m)} = [\![1 \quad 1 \quad 1 \quad \ldots \quad 1 \quad 1 \quad 1]\!]_{m \times m} \qquad (8.3)$$

For n times repeated integration, the related operational matrix $\mathbf{H_{n(m)}}$ can be obtained as

$$\mathbf{H_{n(m)}} = \frac{h^{n-1}}{(n-1)!}[\![0 \quad 1^{n-1} \quad 2^{n-1} \quad \ldots (m-1)^{n-1}]\!]_{m \times m}$$
$$(8.4)$$

Using equations (8.3) and (8.4), the transfer function of any plant may be expressed in the delta operational transfer function (DOTF) form.

For example, consider a plant with the transfer function $G(s)$ given by $G(s) = (s + a)^{-1}, a > 0$ and its impulse response $g(t) = \exp(-at), t \geq 0$.

We can write

$$G(s) = a^{-1}[as^{-1} - a^2 s^{-2} + a^3 s^{-3} - \cdots \infty] \qquad (8.5)$$

By replacing integrators of different orders in equation (8.5) by their operational equivalents using equation (8.4), we have

$$\mathbf{DOTF} = a^{-1}[\![a\mathbf{H1}_{(m)} - a^2\mathbf{H2}_{(m)} + a^3\mathbf{H3}_{(m)} - \cdots \infty]\!]_{m \times m}$$
$$(8.6)$$

Upon simplification, we get

$$\mathbf{DOTF} = [\![1 \quad \exp(-ah) \quad \exp(-2ah) \quad \cdots \quad \exp[-(m-1)ah]]\!]_{m \times m}$$
$$(8.7)$$

where the elements of the first row of the upper triangular matrix **DOTF** are the samples of impulse response $g(t)$ with a sampling interval of h s.

For a step input, $r1(t) = u(t)$. Then expressing $r1(t)$ in the delta function domain, using equation (8.1), we have

$$r1(t) = [1 \quad 1 \quad 1 \quad \cdots \quad 1 \quad 1]_{1 \times m} \boldsymbol{\Delta}_{(\mathbf{m})}(\mathbf{t}) \triangleq \mathbf{R1} \cdot \boldsymbol{\Delta}_{(\mathbf{m})}(\mathbf{t})$$

For this input and the first-order plant described by equation (8.7), the output $c1(t)$ can be expressed as

$$\mathbf{C1} = \mathbf{R1} \cdot \mathbf{DOTF} \qquad (8.8)$$

where, $\mathbf{C1}$ is the output vector in the DF domain.

8.1.2 Sample-and-hold function domain operational technique [5]

Sample-and-hold functions were defined in Section 2.3. This orthogonal function set is a variant of the block pulse function set and is an effective tool for the analysis of sample-and-hold systems. Figure 8.2 shows a set of sample-and-hold (SHF) functions.

It was shown by Deb et al. [5] that the operational matrix for integration of order m in SHF domain is expressed by

$$\mathbf{P1}_{(\mathbf{m})} = h [\![0 \quad 1 \quad 1 \quad \cdots \quad 1 \quad 1 \quad 1]\!]_{m \times m} \qquad (8.9)$$

For n times repeated integration, the related operational matrix $\mathbf{P}_{\mathbf{n}_{(\mathbf{m})}}$ is obtained as [5]

$$\mathbf{P}_{\mathbf{n}_{(\mathbf{m})}} = \frac{h^n}{n!} [\![0 \quad (1^n - 0^n) \quad (2^n - 1^n) \quad \cdots$$
$$[(m-1)^n - (m-2)^n]]\!]_{m \times m} \quad \text{where } m \geq 2 \quad (8.10)$$

Using equation (8.10), the transfer function of any plant may be expressed in the sample-and-hold operational transfer function (SHOTF) form.

Similar to the DF domain technique, SHOTF for the plant $G(s) = (s + a)^{-1}$ can be derived by expanding $G(s)$ binomially and replacing integrators of different orders by its SHF operational equivalents using equation (8.10). Thus,

$$\mathbf{SHOTF} = a^{-1} [\![a\mathbf{P1}_{(\mathbf{m})} - a^2\mathbf{P2}_{(\mathbf{m})} + a^3\mathbf{P3}_{(\mathbf{m})} - \cdots \infty]\!]_{m \times m}$$
$$(8.11)$$

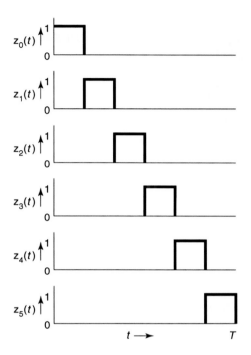

Figure 8.2. *A set of sample-and-hold functions.*

Upon simplification, we get

$$\mathbf{SHOTF} = \frac{1 - \exp(-ah)}{a}$$
$$\times [\![0 \quad 1 \quad \exp(-ah) \quad \dots \quad \exp[-(m-2)ah]]\!]_{m \times m} \quad (8.12)$$

For a step input, $r2(t) = u(t)$. Then expressing $r2(t)$ in the SHF domain, using equation (2.26), we have

$$r2(t) = [1 \quad 1 \quad 1 \quad \dots \quad 1 \quad 1]_{1 \times m} \mathbf{Z_{(m)}(t)} \triangleq \mathbf{R2 \cdot Z_{(m)}(t)}$$

For this input and the first-order plant described by equation (8.12), the output $c2(t)$ can be expressed as

$$\mathbf{C2 = R2 \cdot SHOTF} \quad (8.13)$$

where, **C2** is the output vector in the SHF domain.

8.2 Microprocessor Based Simulation of Linear Single-Input Single-Output (SISO) Sampled-Data Systems [7]

The basic philosophy of microprocessor based simulation of a linear SISO sampled-data system is to configure the system with the help of a program. This is done in such a fashion that when samples of the input signal are accessed by the microprocessor, the program accepts those sample values and comes up with relevant output of the system.

At first, impulse response $g1(t)$ (say) of the system transfer function $G1(s)$ obtained and samples of $g1(t)$, e.g., $g1_0, g1_1, g1_2, \ldots$, etc., taken at equal intervals of h s, are stored in appropriate memory locations of the microprocessor. These sample values are utilized for constructing the relevant operational transfer functions of the specified system.

8.2.1 Delta domain technique for linear SISO sampled-data systems

This technique is applicable to sampled-data systems without any hold device. Basically, equations (8.1) and (8.7) are used to arrive at the desired output.

Using equation (8.1), any input function $r3(t)$ may be expressed in terms an m-term delta function set as

$$r3(t) \approx [r3_0 \ r3_1 \ \ldots \ r3_{(m-1)}]\mathbf{\Delta_{(m)}(t)} \triangleq \mathbf{R3} \cdot \mathbf{\Delta_{(m)}(t)} \quad (8.14)$$

By virtue of equation (8.7), the DOTF of the system is given by

$$\mathbf{DOTF1} = [\![g1_0 \ g1_1 \ g1_2 \ \ldots \ g1_i, \ \ldots \ g1_{(m-1)}]\!]_{m \times m} \quad (8.15)$$

(i) Open-loop system

For an open-loop system shown in Fig. 8.3, following equation (8.8), using equations (8.14) and (8.15), output of the system is

$$\mathbf{C3} = \mathbf{R3} \cdot \mathbf{DOTF1} \quad (8.16)$$

Figure 8.3. *An open-loop system with delta operational transfer function* **DOTF1**.

If **C3** is expressed as

$$\mathbf{C3} \triangleq [c3_0 \quad c3_1 \quad c3_2 \quad \ldots \quad c3_i \quad \ldots \quad c3_{(m-1)}]$$

then we have from equations (8.14)–(8.16)

$$c3_i = \sum_{j=0}^{i} g1_j r3_{(i-J)} \tag{8.17}$$

We consider a step function as input, and hence

$$r3_0 = r3_1 = r3_2 = \cdots = r3_i = \cdots = r3_{(m-1)} = 1$$

Then, from equation (8.17)

$$c3_i = \sum_{j=0}^{i} g1_j \tag{8.18}$$

Hence, knowing $g1_i$'s and $r3_i$'s, the samples of the output $c3_i$'s can be computed.

Illustration

Consider that $G2(s) = (s + 1)^{-1}$

Then, for $h = 1/16$ s, the DOTF for the system, by virtue of equation (8.7), is

$$\mathbf{DOTF2} = [\![64 \quad 5E \quad 58 \quad 53 \quad 4E \quad 49 \quad 44 \quad 40 \quad 3C \quad 38 \quad 35$$
$$32 \quad 2F \quad 2C \quad 29 \quad 26 \quad 24 \quad 22 \quad 20 \quad 1E]\!]_{20 \times 20}$$
$$\tag{8.19}$$

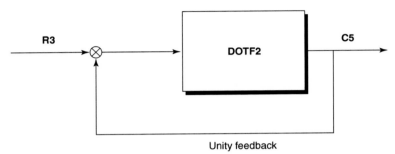

Figure 8.4. *A closed-loop system with delta operational transfer function* **DOTF2** *and unity feedback.*

Here, h is chosen to be equal to $1/16$ s because of convenience in calculating the hex point. Also, the samples of $g2(t)$ are multiplied by 100 (decimal) for decreasing the error in base conversion. The fractional part after multiplication has been rounded off.

For a step input, using equations (8.18) and (8.19), the output of the system **C4** is given by

$$\mathbf{C4} = [64 \quad C2 \quad 11A \quad 16D \quad 1BB \quad 204 \quad 248 \quad 288$$
$$2C4 \quad 2FC \quad 331 \quad 363 \quad 392 \quad 3BE \quad 3E7$$
$$40D \quad 431 \quad 453 \quad 473 \quad 491]_{1\times 20} \qquad (8.20)$$

Obviously, the final result is to be scaled by $1/100$ to get the actual result that clearly matches the computed output of the system.

(ii) Closed-loop system

Let a unity feedback path be provided to the open-loop system described earlier. The closed-loop control system thus obtained will have an equivalent open-loop DOTF (say, **DOTF3**) given as (Fig. 8.4)

$$\mathbf{DOTF3} = \mathbf{DOTF2}[\mathbf{I} + \mathbf{DOTF2}]^{-1} \qquad (8.21)$$

For a step input **R3**, the output **C5** of the system may be obtained using equations (8.16) and (8.21) in a similar manner. That is

$$\mathbf{C5} = \mathbf{R3} \cdot \mathbf{DOTF3}$$

However, it is evident that computation of the inverse of equation (8.21), though not impossible, is not attractive in a microprocessor. Hence, a recursive method is undertaken to compute **C5**. It is noted that $c5_0$, the first element of **C5** is

$$c5_0 = \frac{r3_0 g3_0}{1 + g3_0}$$

where, $r3_0$ and $g3_0$ are the first elements of **R3** and **DOTF3**, respectively, and its ith element is given by

$$c5_i = \frac{1}{(1 + g3_0)} \sum_{p=1}^{i} [r3_{(p-1)} - c3_{(p-1)}] g3_{|i-(p-1)|} + r3_i g3_0$$

where $i \geq 1$ and $r3_i$, is the ith element of **R3**; $g3_i$ is the ith element of **DOTF3**.

The algorithm used for the computation is outlined below.

(i) Subtract the elements of **C5** from the respective elements of **R3** starting from the first element (i.e., $p = 1$) up to the ith element.

(ii) Store the results $[r3_{(p-1)} - c3_{(p-1)}] = r3'_{(p-1)}$, say, in a new vector **R3′** in i (i.e., for $p = 1$ to i) locations.

(iii) Find the products $g3_0 r3_i$ and $r3'_{(p-1)} g3_{|i-(p-1)|}$ for $p = 1$ to i, and store in $(i + 1)$ memory locations.

(iv) Find the sum of the products stored in step (iii).

(v) Find $(1 + g3_0)$.

(vi) Divide the result obtained in step (iv) by the result of step (v) to get $c5_i$.

(vii) Stop.

Illustration

Consider the system $G2(s) = (s + 1)^{-1}$ with a unity feedback loop.

By conventional z-transform analysis, with sampling period $h = 1/16\,\text{s}$, the output $c5(z)$ of this system is obtained as

$$c5(z) = \frac{z^2}{z^2 - 1.88249z + 0.88249}$$

For the DOTF analysis of this system, we proceed in a manner similar to the open-loop system described earlier. The DOTF for the closed-loop system is derived as

$$\mathbf{DOTF3} = [\![64\ 58\ 4E\ 45\ 3D\ 36\ 2F\ 2A\ 25\ 20]\!]_{10\times10} \quad (8.22)$$

For a step input, using equations (8.18) and (8.22), the output of the system **C5** is given by

$$\mathbf{C5} = [64\ BC\ 10A\ 14F\ 18B\ 1C1\ 1F0\ 21A\ 23F\ 25F]_{1\times10}$$

Obviously, the final result is scaled by $1/100$ to get the actual result that clearly matches the z-transform output of the system.

8.2.2 Sampled-and-hold domain technique for linear SISO sampled-data system

Inclusion of a zero-order hold (ZOH) in the system $G1(s)$ introduces an additional transfer function $G0(s) = [1 - \exp(-hs)]/s$, where h is the sampling period. The essence of this transfer function can be interpreted in the Laplace domain as the difference of integration of the system transfer function and its integration delayed by one sampling period. Its SHF domain equivalent may be obtained by application of the following algorithm:

(i) Find impulse response $g1(t)$ from the Laplace domain transfer function $G1(s)$.
(ii) Integrate $g1(t)$ between limits 0 and t to obtain $G1(t)$.
(iii) Take samples of $G1(t)$ at intervals of h s.
(iv) Delay the above sample sequence by one sampling period and subtract each element from the corresponding element of above sequence.
(v) Stop.

Thus, in step (iv) we get the impulse response $g1'(t)$ of the cascaded system in the SHF domain.

Hence, if a ZOH is cascaded with a transfer function $G1(s)$, then with the help of above interpretation, we can convert the transfer function to its operational form, which includes the effect of a ZOH, in the following manner.

Figure 8.5. *An open-loop system with sample-and-hold operational transfer function* **SHOTF1**.

The impulse response of the cascaded system in the SHF domain may be expressed as

$$g1'(t) = [G1(0) \quad \{G1(h) - G1(0)\} \quad \{G1(2h) - G1(h)\}$$
$$\{G1(3h) - G1(2h)\} \quad \cdots$$
$$\{G1[(m-1)h] - G1[(m-2)h]\}]$$

Using $g1'(t)$, the SHOTF for the system may be obtained as

$$\textbf{SHOTF1} = [\![G1(0) \quad \{G1(h) - G1(0)\} \quad \{G1(2h) - G1(h)\}$$
$$\{G1(3h) - G1(2h)\} \quad \cdots$$
$$\{G1[(m-1)h - G1][(m-2)h]\}]\!]_{m \times m} \quad (8.23)$$

For the system having $G(s) = (s + a)^{-1}$, **SHOTF1**, after simplification, turns out to be equal to the **SHOTF** of equation (8.23).

Using equation (2.26), any input function $r4(t)$ may be expressed in terms of an m-term SHF function set as

$$r4(t) \approx [r4_0 \ r4_1 \ \cdots \ r4_{(m-1)}]\textbf{Z}_{(\textbf{m})}(t) \triangleq \textbf{R4} \cdot \textbf{Z}_{(\textbf{m})}(t) \quad (8.24)$$

(i) Open-loop system
Using equations (8.23) and (8.24), the output for an open-loop system can be written as (Fig. 8.5)

$$\textbf{C6} = \textbf{R4} \cdot \textbf{SHOTF1} \quad (8.25)$$

Similar to the DOTF analysis, we can express ith element of **C6** as

$$c6_i = \sum_{j=0}^{i} g1'_j r4_{(i-j)} \qquad (8.26)$$

If **R4** is a step input, then we have

$$c6_i = \sum_{j=0}^{i} g1'_j \qquad (8.27)$$

Illustration

For the system having transfer function $G2(s)$, its sample-and-hold operational transfer function may be obtained from equation (8.23) as

$$\mathbf{SHOTF1} = [\![0 \ 6 \ 6 \ 5 \ 5 \ 5 \ 4 \ 4 \ 4 \ 4 \ 3$$
$$3 \ 3 \ 3 \ 3 \ 3 \ 2 \ 2 \ 2 \ 2]\!]_{20 \times 20} \qquad (8.28)$$

where 20 basis functions have been considered and the sampling period $h = 1/16$ s. For a step input, the output is given by

$$\mathbf{C6} = [0 \ 6 \ C \ 11 \ 16 \ 1B \ 1F \ 23 \ 27 \ 2B \ 2E$$
$$32 \ 35 \ 38 \ 3A \ 3D \ 3F \ 41 \ 44 \ 45]_{1 \times 20} \qquad (8.29)$$

which, when scaled properly, matches the actual output very closely.

(ii) Closed-loop system

Analogous to the DOTF analysis, equivalent open-loop SHOTF (say, **SHOTF2**) for the above system with unity feedback may be obtained as

$$\mathbf{SHOTF2} = \mathbf{SHOTF1} \left[\mathbf{I} + \mathbf{SHOTF1}\right]^{-1} \qquad (8.30)$$

For a step input **R4**, using equations (8.25) and (8.30), the output **C7** is obtained as

$$\mathbf{C7} = \mathbf{R4} \cdot \mathbf{SHOTF2}$$

Like the DOTF analysis, the ith element of **C7** may be obtained by a similar recursive method using a similar algorithm.

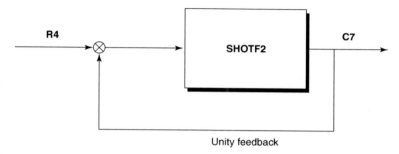

Figure 8.6. *A closed-loop system with sample-and-hold operational transfer function* **SHOTF1** *and unity feedback.*

Illustration

For the system having a transfer function $G2(s) = (s + 1)^{-1}$, the SHOTF is given by equation (8.28). Then, following equation (8.30), the closed-loop SHOTF is obtained as

$$\textbf{SHOTF2} = \llbracket 0 \; 6 \; 5 \; 5 \; 4 \; 4 \; 3 \; 3 \; 2 \; 2 \; 2$$
$$2 \; 1 \; 1 \; 1 \; 1 \; 1 \; 1 \; 1 \; 1 \rrbracket_{20 \times 20} \quad (8.31)$$

For a step input **R4**, the output **C7** is obtained as

$$\textbf{C7} = \begin{bmatrix} 0 & 6 & B & 10 & 14 & 18 & 1B & 1E & 20 & 22 & 24 \\ & 26 & 27 & 29 & 2A & 2B & 2C & 2C & 2D & 2E \end{bmatrix}_{1 \times 20} \quad (8.32)$$

The result, after required scaling, closely matches the exact solution derived from z-transform analysis.

8.3 Conclusion

In the foregoing, Dirac delta function set and sample-and-hold function set have been used for microprocessor-based simulation of discrete time as well as sample-and-hold systems. This implies that such simulations could be useful for identification of control systems with known input-output sequence.

A few open-loop and closed-loop sampled-data systems-with or without hold device-have been studied with the help of the developed algorithm and the results obtained are compared with

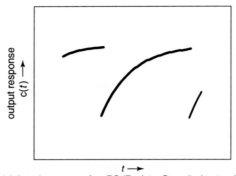

(a) Actual response of an RC (Resistor-Capacitor) network

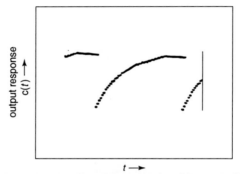

(b) Microprocessor based simulation of the RC network of (a)

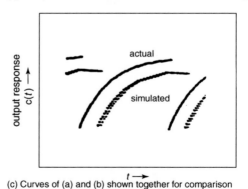

(c) Curves of (a) and (b) shown together for comparison

Figure 8.7. *Step response of an open-loop first-order system having a transfer function* $G(s) = (s + 1)^{-1}$ *with sample-and-hold [all figures scanned from cathode ray oscilloscope (CRO) response].*

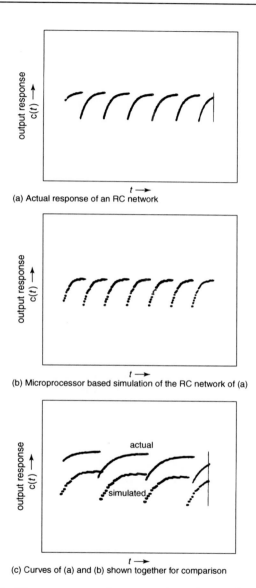

(a) Actual response of an RC network

(b) Microprocessor based simulation of the RC network of (a)

(c) Curves of (a) and (b) shown together for comparison

Figure 8.8. *Step response of an open-loop first-order system having a transfer function* $G(s) = (s + 1)^{-1}$ *with sample-and-hold and unity feedback [all figures scanned from cathode ray oscilloscope (CRO) response].*

exact solutions. It is found that the results match very nicely with the exact solutions. However, performing mathematical operations in the microprocessor were a bit inconvenient. For sampled-data system analysis, the outputs could not be visualized in a cathode ray oscilloscope (CRO). They were observed only as stored data form. This last difficulty was absent in case of sample-and-hold systems and a few sample photographs are shown in Figs. 8.7 and 8.8. Such simulation is also possible for continuous time systems using the newly proposed triangular function. Work is in progress to analyze and identify simulated systems via this function set.

References

1. Jiang, J. H. and Schaufelberger, W., *Block pulse functions and their applications in control systems*, LNCIS-179, Springer-Verlag, Berlin, 1992.
2. Deb, A., Sarkar, G. and Sen, S. K., Block pulse functions, the most fundamental of all piecewise constant basis functions, Int. J. Syst. Sci., vol. **25**, no. 2, pp. 351–363, 1994.
3. Deb, A., Sarkar, G., Bhattacharjee, M. and Sen, S. K., All-integrator approach to linear SISO control system analysis using block pulse functions (BPF), J. Franklin Instt., vol. **334B**, no. 2, pp. 319–335, 1997.
4. Deb, A., Sarkar, G. and Dasgupta, Anindita, A complementary pair of orthogonal triangular function sets and its application to the analysis of SISO control systems, J. Instt. Engrs (India), vol. **84**, December, pp. 120–129, 2003.
5. Deb, A., Sarkar, G., Bhattacharjee, M. and Sen, S. K., A new set of piecewise constant orthogonal functions for the analysis of linear SISO systems with sample-and-hold, J. Franklin Instt, vol. **335B**, no. 2, pp. 333–358, 1998.
6. Deb, A., Sarkar, G., Bhattacharjee, M. and Sen, S. K., Analysis of linear discrete SISO control systems via a set of delta functions, IEE Proc. Control Theory and Appl., vol. **143**, no. 6, pp. 514–518, 1996.
7. Sarkar, G., Dasgupta, A. and Deb, A., Microprocessor based simulation of sampled data systems with/without a hold device using a set of sample-and-hold functions and Dirac delta functions, J. Franklin Instt. vol. **342**, no. 1, pp. 85–95. 2005.

Index

Lightning Source UK Ltd.
Milton Keynes UK
R1183800001B/R11838PG176124UKX1B/1/P